Iterative Methods for Linear and Nonlinear Equations

Frontiers in Applied Mathematics

Frontiers in Applied Mathematics is a series that presents new mathematical or computational approaches to significant scientific problems. Beginning with Volume 4, the series reflects a change in both philosophy and format. Each volume focuses on a broad application of general interest to applied mathematicians as well as engineers and other scientists.

This unique series will advance the development of applied mathematics through the rapid publication of short, inexpensive books that lie on the cutting edge of research.

Frontiers in Applied Mathematics

Iterative Methods for Linear and Nonlinear Equations

C. T. Kelley
North Carolina State University

Society for Industrial and Applied Mathematics
siam.
Philadelphia 1995

Library of Congress Cataloging-in-Publication Data

Kelley, C. T.
    Iterative methods for linear and nonlinear equations / C. T.
    Kelley
        p. cm. — (Frontiers in applied mathematics ; vol. 16)
        Includes bibliographical references and index.
        ISBN 0-89871-352-8 (pbk.)
        1. Iterative methods (Mathematics) I. Title. II. Series:
Frontiers in applied mathematics ; 16.
QA297.8.K45 1995
519.4—dc20   95-32249

To Polly H. Thomas, 1906-1994, devoted mother and grandmother

Contents

Preface

This book on iterative methods for linear and nonlinear equations can be used as a tutorial and a reference by anyone who needs to solve nonlinear systems of equations or large linear systems. It may also be used as a textbook for introductory courses in nonlinear equations or iterative methods or as source material for an introductory course in numerical analysis at the graduate level. We assume that the reader is familiar with elementary numerical analysis, linear algebra, and the central ideas of direct methods for the numerical solution of dense linear systems as described in standard texts such as [7], [105], or [184].

Our approach is to focus on a small number of methods and treat them in depth. Though this book is written in a finite-dimensional setting, we have selected for coverage mostly algorithms and methods of analysis which extend directly to the infinite-dimensional case and whose convergence can be thoroughly analyzed. For example, the matrix-free formulation and analysis for GMRES and conjugate gradient is almost unchanged in an infinite-dimensional setting. The analysis of Broyden's method presented in Chapter 7 and the implementations presented in Chapters 7 and 8 are different from the classical ones and also extend directly to an infinite-dimensional setting. The computational examples and exercises focus on discretizations of infinite-dimensional problems such as integral and differential equations.

We present a limited number of computational examples. These examples are intended to provide results that can be used to validate the reader's own implementations and to give a sense of how the algorithms perform. The examples are not designed to give a complete picture of performance or to be a suite of test problems.

The computational examples in this book were done with MATLAB® (version 4.0a on various SUN SPARCstations and version 4.1 on an Apple Macintosh Powerbook 180) and the MATLAB environment is an excellent one for getting experience with the algorithms, for doing the exercises, and for small-to-medium scale production work.[1] MATLAB codes for many of the algorithms are available by anonymous ftp. A good introduction to the latest

---

[1]MATLAB is a registered trademark of The MathWorks, Inc.

version (version 4.2) of MATLAB is the MATLAB Primer [178]; [43] is also a useful resource. If the reader has no access to MATLAB or will be solving very large problems, the general algorithmic descriptions or even the MATLAB codes can easily be translated to another language.

Parts of this book are based upon work supported by the National Science Foundation and the Air Force Office of Scientific Research over several years, most recently under National Science Foundation Grant Nos. DMS-9024622 and DMS-9321938. Any opinions, findings, and conclusions or recommendations expressed in this material are those of the author and do not necessarily reflect the views of the National Science Foundation or of the Air Force Office of Scientific Research.

Many of my students and colleagues discussed various aspects of this project with me and provided important corrections, ideas, suggestions, and pointers to the literature. I am especially indebted to Jim Banoczi, Jeff Butera, Steve Campbell, Tony Choi, Moody Chu, Howard Elman, Jim Epperson, Andreas Griewank, Laura Helfrich, Ilse Ipsen, Lea Jenkins, Vickie Kearn, Belinda King, Debbie Lockhart, Carl Meyer, Casey Miller, Ekkehard Sachs, Jeff Scroggs, Joseph Skudlarek, Mike Tocci, Gordon Wade, Homer Walker, Steve Wright, Zhaqing Xue, Yue Zhang, and an anonymous reviewer for their contributions and encouragement.

Most importantly, I thank Chung-Wei Ng and my parents for over one hundred and ten years of patience and support.

C. T. Kelley
Raleigh, North Carolina
January, 1998

How to get the software

A collection of MATLAB codes has been written to accompany this book. The MATLAB codes can be obtained by anonymous ftp from the MathWorks server `ftp.mathworks.com` in the directory `pub/books/kelley`, from the MathWorks World Wide Web site,

    `http://www.mathworks.com`

or from SIAM's World Wide Web site

    `http://www.siam.org/books/kelley/kelley.html`

One can obtain MATLAB from
The MathWorks, Inc.
24 Prime Park Way
Natick, MA 01760,
Phone: (508) 653-1415
Fax: (508) 653-2997
E-mail: info@mathworks.com
WWW: http://www.mathworks.com

Linear Equations

Chapter 1

Basic Concepts and Stationary Iterative Methods

1.1. Review and notation

We begin by setting notation and reviewing some ideas from numerical linear algebra that we expect the reader to be familiar with. An excellent reference for the basic ideas of numerical linear algebra and direct methods for linear equations is [184].

We will write linear equations as

(1.1) $$Ax = b,$$

where A is a nonsingular $N \times N$ matrix, $b \in R^N$ is given, and

$$x^* = A^{-1}b \in R^N$$

is to be found.

Throughout this chapter x will denote a potential solution and $\{x_k\}_{k \geq 0}$ the sequence of iterates. We will denote the ith component of a vector x by $(x)_i$ (note the parentheses) and the ith component of x_k by $(x_k)_i$. We will rarely need to refer to individual components of vectors.

In this chapter $\|\cdot\|$ will denote a norm on R^N as well as the *induced matrix norm*.

DEFINITION 1.1.1. *Let $\|\cdot\|$ be a norm on R^N. The* induced matrix norm *of an $N \times N$ matrix A is defined by*

$$\|A\| = \max_{\|x\|=1} \|Ax\|.$$

Induced norms have the important property that

$$\|Ax\| \leq \|A\|\|x\|.$$

Recall that the *condition number* of A relative to the norm $\|\cdot\|$ is

$$\kappa(A) = \|A\|\|A^{-1}\|,$$

where $\kappa(A)$ is understood to be infinite if A is singular. If $\|\cdot\|$ is the l^p norm

$$\|x\|_p = \left(\sum_{j=1}^{N} |(x)_i|^p \right)^{1/p}$$

we will write the condition number as κ_p.

Most iterative methods terminate when the residual

$$r = b - Ax$$

is sufficiently small. One termination criterion is

(1.2)
$$\frac{\|r_k\|}{\|r_0\|} < \tau,$$

which can be related to the error

$$e = x - x^*$$

in terms of the condition number.

LEMMA 1.1.1. *Let $b, x, x_0 \in R^N$. Let A be nonsingular and let $x^* = A^{-1}b$.*

(1.3)
$$\frac{\|e\|}{\|e_0\|} \le \kappa(A) \frac{\|r\|}{\|r_0\|}.$$

Proof. Since

$$r = b - Ax = -Ae$$

we have

$$\|e\| = \|A^{-1}Ae\| \le \|A^{-1}\|\|Ae\| = \|A^{-1}\|\|r\|$$

and

$$\|r_0\| = \|Ae_0\| \le \|A\|\|e_0\|.$$

Hence

$$\frac{\|e\|}{\|e_0\|} \le \frac{\|A^{-1}\|\|r\|}{\|A\|^{-1}\|r_0\|} = \kappa(A) \frac{\|r\|}{\|r_0\|},$$

as asserted. □

The termination criterion (1.2) depends on the initial iterate and may result in unnecessary work when the initial iterate is good and a poor result when the initial iterate is far from the solution. For this reason we prefer to terminate the iteration when

(1.4)
$$\frac{\|r_k\|}{\|b\|} < \tau.$$

The two conditions (1.2) and (1.4) are the same when $x_0 = 0$, which is a common choice, particularly when the linear iteration is being used as part of a nonlinear solver.

1.2. The Banach Lemma and approximate inverses

The most straightforward approach to an iterative solution of a linear system is to rewrite (1.1) as a linear fixed-point iteration. One way to do this is to write $Ax = b$ as

$$(1.5) \qquad x = (I - A)x + b,$$

and to define the *Richardson iteration*

$$(1.6) \qquad x_{k+1} = (I - A)x_k + b.$$

We will discuss more general methods in which $\{x_k\}$ is given by

$$(1.7) \qquad x_{k+1} = Mx_k + c.$$

In (1.7) M is an $N \times N$ matrix called the *iteration matrix*. Iterative methods of this form are called *stationary iterative methods* because the transition from x_k to x_{k+1} does not depend on the history of the iteration. The Krylov methods discussed in Chapters 2 and 3 are not stationary iterative methods.

All our results are based on the following lemma.

LEMMA 1.2.1. *If M is an $N \times N$ matrix with $\|M\| < 1$ then $I - M$ is nonsingular and*

$$(1.8) \qquad \|(I - M)^{-1}\| \le \frac{1}{1 - \|M\|}.$$

Proof. We will show that $I - M$ is nonsingular and that (1.8) holds by showing that the series

$$\sum_{l=0}^{\infty} M^l = (I - M)^{-1}.$$

The partial sums

$$S_k = \sum_{l=0}^{k} M^l$$

form a Cauchy sequence in $R^{N \times N}$. To see this note that for all $m > k$

$$\|S_k - S_m\| \le \sum_{l=k+1}^{m} \|M^l\|.$$

Now, $\|M^l\| \le \|M\|^l$ because $\|\cdot\|$ is an matrix norm that is induced by a vector norm. Hence

$$\|S_k - S_m\| \le \sum_{l=k+1}^{m} \|M\|^l = \|M\|^{k+1} \left(\frac{1 - \|M\|^{m-k}}{1 - \|M\|} \right) \to 0$$

as $m, k \to \infty$. Hence the sequence S_k converges, say to S. Since $MS_k + I = S_{k+1}$, we must have $MS + I = S$ and hence $(I - M)S = I$. This proves that $I - M$ is nonsingular and that $S = (I - M)^{-1}$.

Noting that

$$\|(I - M)^{-1}\| \le \sum_{l=0}^{\infty} \|M\|^l = (1 - \|M\|)^{-1}.$$

proves (1.8) and completes the proof. $\square$

The following corollary is a direct consequence of Lemma 1.2.1.

COROLLARY 1.2.1. *If* $\|M\| < 1$ *then the iteration* (1.7) *converges to* $x = (I - M)^{-1}c$ *for all initial iterates* x_0.

A consequence of Corollary 1.2.1 is that Richardson iteration (1.6) will converge if $\|I - A\| < 1$. It is sometimes possible to *precondition* a linear equation by multiplying both sides of (1.1) by a matrix B

$$BAx = Bb$$

so that convergence of iterative methods is improved. In the context of Richardson iteration, the matrices B that allow us to apply the Banach lemma and its corollary are called *approximate inverses*.

DEFINITION 1.2.1. B *is an* approximate inverse *of* A *if* $\|I - BA\| < 1$.

The following theorem is often referred to as the *Banach Lemma*.

THEOREM 1.2.1. *If* A *and* B *are* $N \times N$ *matrices and* B *is an approximate inverse of* A, *then* A *and* B *are both nonsingular and*

(1.9) $\|A^{-1}\| \le \dfrac{\|B\|}{1 - \|I - BA\|}, \quad \|B^{-1}\| \le \dfrac{\|A\|}{1 - \|I - BA\|},$

and

(1.10) $\|A^{-1} - B\| \le \dfrac{\|B\|\|I - BA\|}{1 - \|I - BA\|}, \quad \|A - B^{-1}\| \le \dfrac{\|A\|\|I - BA\|}{1 - \|I - BA\|}.$

Proof. Let $M = I - BA$. By Lemma 1.2.1 $I - M = I - (I - BA) = BA$ is nonsingular. Hence both A and B are nonsingular. By (1.8)

(1.11) $\|A^{-1}B^{-1}\| = \|(I - M)^{-1}\| \le \dfrac{1}{1 - \|M\|} = \dfrac{1}{1 - \|I - BA\|}.$

Since $A^{-1} = (I - M)^{-1}B$, inequality (1.11) implies the first part of (1.9). The second part follows in a similar way from $B^{-1} = A(I - M)^{-1}$.

To complete the proof note that

$$A^{-1} - B = (I - BA)A^{-1}, A - B^{-1} = B^{-1}(I - BA),$$

and use (1.9). $\square$

Richardson iteration, preconditioned with approximate inversion, has the form

(1.12) $x_{k+1} = (I - BA)x_k + Bb.$

If the norm of $I - BA$ is small, then not only will the iteration converge rapidly, but, as Lemma 1.1.1 indicates, termination decisions based on the preconditioned residual $Bb - BAx$ will better reflect the actual error. This method is a very effective technique for solving differential equations, integral equations, and related problems [15], [6], [100], [117], [111]. Multigrid methods [19], [99], [126], can also be interpreted in this light. We mention one other approach, *polynomial preconditioning*, which tries to approximate A^{-1} by a polynomial in A [123], [179], [169].

1.3. The spectral radius

The analysis in § 1.2 related convergence of the iteration (1.7) to the norm of the matrix M. However the norm of M could be small in some norms and quite large in others. Hence the performance of the iteration is not completely described by $\|M\|$. The concept of spectral radius allows us to make a complete description.

We let $\sigma(A)$ denote the set of eigenvalues of A.

DEFINITION 1.3.1. *The spectral radius of an $N \times N$ matrix A is*

$$(1.13) \qquad \rho(A) = \max_{\lambda \in \sigma(A)} |\lambda| = \lim_{n \to \infty} \|A^n\|^{1/n}.$$

The term on the right-hand side of the second equality in (1.13) is the limit used by the radical test for convergence of the series $\sum A^n$.

The spectral radius of M is independent of any particular matrix norm of M. It is clear, in fact, that

$$(1.14) \qquad \rho(A) \le \|A\|$$

for any induced matrix norm. The inequality (1.14) has a partial converse that allows us to completely describe the performance of iteration (1.7) in terms of spectral radius. We state that converse as a theorem and refer to [105] for a proof.

THEOREM 1.3.1. *Let A be an $N \times N$ matrix. Then for any $\epsilon > 0$ there is a norm $\| \cdot \|$ on R^N such that*

$$\rho(A) > \|A\| - \epsilon.$$

A consequence of Theorem 1.3.1, Lemma 1.2.1, and Exercise 1.5.1 is a characterization of convergent stationary iterative methods. The proof is left as an exercise.

THEOREM 1.3.2. *Let M be an $N \times N$ matrix. The iteration (1.7) converges for all $c \in R^N$ if and only if $\rho(M) < 1$.*

1.4. Matrix splittings and classical stationary iterative methods

There are ways to convert $Ax = b$ to a linear fixed-point iteration that are different from (1.5). Methods such as Jacobi, Gauss–Seidel, and sucessive overrelaxation (SOR) iteration are based on *splittings* of A of the form

$$A = A_1 + A_2,$$

where A_1 is a nonsingular matrix constructed so that equations with A_1 as coefficient matrix are easy to solve. Then $Ax = b$ is converted to the fixed-point problem

$$x = A_1^{-1}(b - A_2 x).$$

The analysis of the method is based on an estimation of the spectral radius of the iteration matrix $M = -A_1^{-1}A_2$.

For a detailed description of the classical stationary iterative methods the reader may consult [89], [105], [144], [193], or [200]. These methods are usually less efficient than the Krylov methods discussed in Chapters 2 and 3 or the more modern stationary methods based on multigrid ideas. However the classical methods have a role as preconditioners. The limited description in this section is intended as a review that will set some notation to be used later.

As a first example we consider the Jacobi iteration that uses the splitting

$$A_1 = D, A_2 = L + U,$$

where D is the diagonal of A and L and U are the (strict) lower and upper triangular parts. This leads to the iteration matrix

$$M_{JAC} = -D^{-1}(L + U).$$

Letting $(x_k)_i$ denote the ith component of the kth iterate we can express Jacobi iteration concretely as

$$(1.15) \qquad (x_{k+1})_i = a_{ii}^{-1}\left(b_i - \sum_{j \neq i} a_{ij}(x_k)_j\right).$$

Note that A_1 is diagonal and hence trivial to invert.

We present only one convergence result for the classical stationary iterative methods.

THEOREM 1.4.1. *Let A be an $N \times N$ matrix and assume that for all $1 \leq i \leq N$*

$$(1.16) \qquad 0 < \sum_{j \neq i} |a_{ij}| < |a_{ii}|.$$

Then A is nonsingular and the Jacobi iteration (1.15) converges to $x^* = A^{-1}b$ for all b.

Proof. Note that the ith row sum of $M = M_{JAC}$ satisfies

$$\sum_{j=1}^{N} |m_{ij}| = \frac{\sum_{j \neq i} |a_{ij}|}{|a_{ii}|} < 1.$$

Hence $\|M_{JAC}\|_\infty < 1$ and the iteration converges to the unique solution of $x = Mx + D^{-1}b$. Also $I - M = D^{-1}A$ is nonsingular and therefore A is nonsingular. $\square$

Gauss–Seidel iteration overwrites the approximate solution with the new value as soon as it is computed. This results in the iteration

$$(x_{k+1})_i = a_{ii}^{-1}\left(b_i - \sum_{j<i} a_{ij}(x_{k+1})_j - \sum_{j>i} a_{ij}(x_k)_j\right),$$

the splitting

$$A_1 = D + L, A_2 = U,$$

and iteration matrix

$$M_{GS} = -(D+L)^{-1}U.$$

Note that A_1 is lower triangular, and hence $A_1^{-1}y$ is easy to compute for vectors y. Note also that, unlike Jacobi iteration, the iteration depends on the ordering of the unknowns. Backward Gauss–Seidel begins the update of x with the Nth coordinate rather than the first, resulting in the splitting

$$A_1 = D + U, A_2 \doteq L,$$

and iteration matrix

$$M_{BGS} = -(D+U)^{-1}L.$$

A symmetric Gauss–Seidel iteration is a forward Gauss–Seidel iteration followed by a backward Gauss–Seidel iteration. This leads to the iteration matrix

$$M_{SGS} = M_{BGS}M_{GS} = (D+U)^{-1}L(D+L)^{-1}U.$$

If A is symmetric then $U = L^T$. In that event

$$M_{SGS} = (D+U)^{-1}L(D+L)^{-1}U = (D+L^T)^{-1}L(D+L)^{-1}L^T.$$

From the point of view of preconditioning, one wants to write the stationary method as a preconditioned Richardson iteration. That means that one wants to find B such that $M = I - BA$ and then use B as an approximate inverse. For the Jacobi iteration,

(1.17) $$B_{JAC} = D^{-1}.$$

For symmetric Gauss–Seidel

(1.18) $$B_{SGS} = (D+L^T)^{-1}D(D+L)^{-1}.$$

The successive overrelaxation iteration modifies Gauss–Seidel by adding a *relaxation parameter* ω to construct an iteration with iteration matrix

$$M_{SOR} = (D+\omega L)^{-1}((1-\omega)D - \omega U).$$

The performance can be dramatically improved with a good choice of ω but still is not competitive with Krylov methods. A further disadvantage is that the choice of ω is often difficult to make. References [200], [89], [193], [8], and the papers cited therein provide additional reading on this topic.

1.5. Exercises on stationary iterative methods

1.5.1. Show that if $\rho(M) \geq 1$ then there are x_0 and c such that the iteration (1.7) fails to converge.

1.5.2. Prove Theorem 1.3.2.

1.5.3. Verify equality (1.18).

1.5.4. Show that if A is symmetric and positive definite (that is $A^T = A$ and $x^T A x > 0$ for all $x \neq 0$) that B_{SGS} is also symmetric and positive definite.

Chapter 2

Conjugate Gradient Iteration

2.1. Krylov methods and the minimization property

In the following two chapters we describe some of the Krylov space methods for linear equations. Unlike the stationary iterative methods, Krylov methods do not have an iteration matrix. The two such methods that we'll discuss in depth, conjugate gradient and GMRES, minimize, at the kth iteration some measure of error over the affine space

$$x_0 + \mathcal{K}_k,$$

where x_0 is the initial iterate and the kth *Krylov* subspace $\mathcal{K}_k$ is

$$\mathcal{K}_k = \text{span}(r_0, Ar_0, \ldots, A^{k-1}r_0)$$

for $k \geq 1$.

The residual is

$$r = b - Ax.$$

So $\{r_k\}_{k\geq 0}$ will denote the sequence of residuals

$$r_k = b - Ax_k.$$

As in Chapter 1, we assume that A is a nonsingular $N \times N$ matrix and let

$$x^* = A^{-1}b.$$

There are other Krylov methods that are not as well understood as CG or GMRES. Brief descriptions of several of these methods and their properties are in § 3.6, [12], and [78].

The conjugate gradient (CG) iteration was invented in the 1950s [103] as a direct method. It has come into wide use over the last 15 years as an iterative method and has generally superseded the Jacobi–Gauss–Seidel–SOR family of methods.

CG is intended to solve symmetric positive definite (spd) systems. Recall that A is *symmetric* if $A = A^T$ and *positive definite* if

$$x^T Ax > 0 \text{ for all } x \neq 0.$$

11

In this section we assume that A is spd. Since A is spd we may define a norm (you should check that this is a norm) by

$$(2.1) \qquad \|x\|_A = \sqrt{x^T A x}.$$

$\|\cdot\|_A$ is called the A-norm. The development in these notes is different from the classical work and more like the analysis for GMRES and CGNR in [134]. In this section, and in the section on GMRES that follows, we begin with a description of what the algorithm does and the consequences of the minimization property of the iterates. After that we describe termination criterion, performance, preconditioning, and at the very end, the implementation.

The kth iterate x_k of CG minimizes

$$(2.2) \qquad \phi(x) = \frac{1}{2} x^T A x - x^T b$$

over $x_0 + \mathcal{K}_k$.

Note that if $\phi(\tilde{x})$ is the minimal value (in R^N) then

$$\nabla \phi(\tilde{x}) = A\tilde{x} - b = 0$$

and hence $\tilde{x} = x^*$.

Minimizing ϕ over any subset of R^N is the same as minimizing $\|x - x^*\|_A$ over that subset. We state this as a lemma.

LEMMA 2.1.1. *Let $S \subset R^N$. If x_k minimizes ϕ over S then x_k also minimizes $\|x^* - x\|_A = \|r\|_{A^{-1}}$ over S.*

Proof. Note that

$$\|x - x^*\|_A^2 = (x - x^*)^T A(x - x^*) = x^T A x - x^T A x^* - (x^*)^T A x + (x^*)^T A x^*.$$

Since A is symmetric and $Ax^* = b$

$$-x^T A x^* - (x^*)^T A x = -2x^T A x^* = -2x^T b.$$

Therefore

$$\|x - x^*\|_A^2 = 2\phi(x) + (x^*)^T A x^*.$$

Since $(x^*)^T A x^*$ is independent of x, minimizing ϕ is equivalent to minimizing $\|x - x^*\|_A^2$ and hence to minimizing $\|x - x^*\|_A$.

If $e = x - x^*$ then

$$\|e\|_A^2 = e^T A e = (A(x - x^*))^T A^{-1}(A(x - x^*)) = \|b - Ax\|_{A^{-1}}^2$$

and hence the A-norm of the error is also the A^{-1}-norm of the residual. $\square$

We will use this lemma in the particular case that $S = x_0 + \mathcal{K}_k$ for some k.

2.2. Consequences of the minimization property

Lemma 2.1.1 implies that since x_k minimizes ϕ over $x_0 + \mathcal{K}_k$

(2.3) $$\|x^* - x_k\|_A \leq \|x^* - w\|_A$$

for all $w \in x_0 + \mathcal{K}_k$. Since any $w \in x_0 + \mathcal{K}_k$ can be written as

$$w = \sum_{j=0}^{k-1} \gamma_j A^j r_0 + x_0$$

for some coefficients $\{\gamma_j\}$, we can express $x^* - w$ as

$$x^* - w = x^* - x_0 - \sum_{j=0}^{k-1} \gamma_j A^j r_0.$$

Since $Ax^* = b$ we have

$$r_0 = b - Ax_0 = A(x^* - x_0)$$

and therefore

$$x^* - w = x^* - x_0 - \sum_{j=0}^{k-1} \gamma_j A^{j+1}(x^* - x_0) = p(A)(x^* - x_0),$$

where the polynomial

$$p(z) = 1 - \sum_{j=0}^{k-1} \gamma_j z^{j+1}$$

has degree k and satisfies $p(0) = 1$. Hence

(2.4) $$\|x^* - x_k\|_A = \min_{p \in \mathbf{P}_k, p(0)=1} \|p(A)(x^* - x_0)\|_A.$$

In (2.4) $\mathbf{P}_k$ denotes the set of polynomials of degree k.

The spectral theorem for spd matrices asserts that

$$A = U\Lambda U^T,$$

where U is an orthogonal matrix whose columns are the eigenvectors of A and Λ is a diagonal matrix with the positive eigenvalues of A on the diagonal. Since $UU^T = U^T U = I$ by orthogonality of U, we have

$$A^j = U\Lambda^j U^T.$$

Hence

$$p(A) = Up(\Lambda)U^T.$$

Define $A^{1/2} = U\Lambda^{1/2}U^T$ and note that

(2.5) $$\|x\|_A^2 = x^T A x = \|A^{1/2}x\|_2^2.$$

Hence, for any $x \in R^N$ and

$$\|p(A)x\|_A = \|A^{1/2}p(A)x\|_2 \le \|p(A)\|_2\|A^{1/2}x\|_2 \le \|p(A)\|_2\|x\|_A.$$

This, together with (2.4) implies that

$$(2.6) \qquad \|x_k - x^*\|_A \le \|x_0 - x^*\|_A \min_{p \in \mathbf{P}_k, p(0)=1} \max_{z \in \sigma(A)} |p(z)|.$$

Here $\sigma(A)$ is the set of all eigenvalues of A.

The following corollary is an important consequence of (2.6).

COROLLARY 2.2.1. *Let A be spd and let $\{x_k\}$ be the CG iterates. Let k be given and let $\{\bar{p}_k\}$ be any kth degree polynomial such that $\bar{p}_k(0) = 1$. Then*

$$(2.7) \qquad \frac{\|x_k - x^*\|_A}{\|x_0 - x^*\|_A} \le \max_{z \in \sigma(A)} |\bar{p}_k(z)|.$$

We will refer to the polynomial $\bar{p}_k$ as a *residual polynomial* [185].

DEFINITION 2.2.1. *The set of kth degree residual polynomials is*

$$(2.8) \qquad \mathcal{P}_k = \{p \mid p \text{ is a polynomial of degree } k \text{ and } p(0) = 1.\}$$

In specific contexts we try to construct sequences of residual polynomials, based on information on $\sigma(A)$, that make either the middle or the right term in (2.7) easy to evaluate. This leads to an upper estimate for the number of CG iterations required to reduce the A-norm of the error to a given tolerance.

One simple application of (2.7) is to show how the CG algorithm can be viewed as a direct method.

THEOREM 2.2.1. *Let A be spd. Then the CG algorithm will find the solution within N iterations.*

Proof. Let $\{\lambda_i\}_{i=1}^N$ be the eigenvalues of A. As a test polynomial, let

$$\bar{p}(z) = \prod_{i=1}^{N}(\lambda_i - z)/\lambda_i.$$

$\bar{p} \in \mathcal{P}_N$ because $\bar{p}$ has degree N and $\bar{p}(0) = 1$. Hence, by (2.7) and the fact that $\bar{p}$ vanishes on $\sigma(A)$,

$$\|x_N - x^*\|_A \le \|x_0 - x^*\|_A \max_{z \in \sigma(A)} |\bar{p}(z)| = 0. \qquad \square$$

Note that our test polynomial had the eigenvalues of A as its roots. In that way we showed (in the absence of all roundoff error!) that CG terminated in finitely many iterations with the exact solution. This is not as good as it sounds, since in most applications the number of unknowns N is very large, and one cannot afford to perform N iterations. It is best to regard CG as an iterative method. When doing that we seek to terminate the iteration when some specified error tolerance is reached.

In the two examples that follow we look at some other easy consequences of (2.7).

THEOREM 2.2.2. *Let A be spd with eigenvectors $\{u_i\}_{i=1}^N$. Let b be a linear combination of k of the eigenvectors of A*

$$b = \sum_{l=1}^{k} \gamma_l u_{i_l}.$$

Then the CG iteration for $Ax = b$ with $x_0 = 0$ will terminate in at most k iterations.

Proof. Let $\{\lambda_{i_l}\}$ be the eigenvalues of A associated with the eigenvectors $\{u_{i_l}\}_{l=1}^k$. By the spectral theorem

$$x^* = \sum_{l=1}^{k} (\gamma_l/\lambda_{i_l}) u_{i_l}.$$

We use the residual polynomial,

$$\bar{p}(z) = \prod_{l=1}^{k} (\lambda_{i_l} - z)/\lambda_{i_l}.$$

One can easily verify that $\bar{p} \in \mathcal{P}_k$. Moreover, $\bar{p}(\lambda_{i_l}) = 0$ for $1 \leq l \leq k$ and hence

$$\bar{p}(A)x^* = \sum_{l=1}^{k} \bar{p}(\lambda_{i_l})\gamma_l/\lambda_{i_l} u_{i_l} = 0.$$

So, we have by (2.4) and the fact that $x_0 = 0$ that

$$\|x_k - x^*\|_A \leq \|\bar{p}(A)x^*\|_A = 0.$$

This completes the proof. □

If the spectrum of A has fewer than N points, we can use a similar technique to prove the following theorem.

THEOREM 2.2.3. *Let A be spd. Assume that there are exactly $k \leq N$ distinct eigenvalues of A. Then the CG iteration terminates in at most k iterations.*

2.3. Termination of the iteration

In practice we do not run the CG iteration until an exact solution is found, but rather terminate once some criterion has been satisfied. One typical criterion is small (say $\leq \eta$) relative residuals. This means that we terminate the iteration after

(2.9) $$\|b - Ax_k\|_2 \leq \eta\|b\|_2.$$

The error estimates that come from the minimization property, however, are based on (2.7) and therefore estimate the reduction in the relative A-norm of the error.

Our next task is to relate the relative residual in the Euclidean norm to the relative error in the A-norm. We will do this in the next two lemmas and then illustrate the point with an example.

LEMMA 2.3.1. *Let A be spd with eigenvalues $\lambda_1 \geq \lambda_2 \geq \ldots \lambda_N$. Then for all $z \in R^N$,*

(2.10)
$$\|A^{1/2}z\|_2 = \|z\|_A$$

and

(2.11)
$$\lambda_N^{1/2}\|z\|_A \leq \|Az\|_2 \leq \lambda_1^{1/2}\|z\|_A.$$

Proof. Clearly

$$\|z\|_A^2 = z^T A z = (A^{1/2}z)^T(A^{1/2}z) = \|A^{1/2}z\|_2^2$$

which proves (2.10).

Let u_i be a unit eigenvector corresponding to λ_i. We may write $A = U\Lambda U^T$ as

$$Az = \sum_{i=1}^N \lambda_i(u_i^T z)u_i.$$

Hence

$$\lambda_N\|A^{1/2}z\|_2^2 = \lambda_N \sum_{i=1}^N \lambda_i(u_i^T z)^2$$

$$\leq \|Az\|_2^2 = \sum_{i=1}^N \lambda_i^2(u_i^T z)^2$$

$$\leq \lambda_1 \sum_{i=1}^N \lambda_i(u_i^T z)^2 = \lambda_1\|A^{1/2}z\|_2^2.$$

Taking square roots and using (2.10) complete the proof. □

LEMMA 2.3.2.

(2.12)
$$\frac{\|b\|_2}{\|r_0\|_2}\frac{\|b - Ax_k\|_2}{\|b\|_2} = \frac{\|b - Ax_k\|_2}{\|b - Ax_0\|_2} \leq \sqrt{\kappa_2(A)}\frac{\|x_k - x^*\|_A}{\|x^* - x_0\|_A}$$

and

(2.13)
$$\frac{\|b - Ax_k\|_2}{\|b\|_2} \leq \frac{\sqrt{\kappa_2(A)}\|r_0\|_2}{\|b\|_2}\frac{\|x_k - x^*\|_A}{\|x^* - x_0\|_A}.$$

Proof. The equality on the left of (2.12) is clear and (2.13) follows directly from (2.12). To obtain the inequality on the right of (2.12), first recall that if $A = U\Lambda U^T$ is the spectral decomposition of A and we order the eigenvalues such that $\lambda_1 \geq \lambda_2 \geq \ldots \lambda_N > 0$, then $\|A\|_2 = \lambda_1$ and $\|A^{-1}\|_2 = 1/\lambda_N$. So $\kappa_2(A) = \lambda_1/\lambda_N$.

Therefore, using (2.10) and (2.11) twice,

$$\frac{\|b - Ax_k\|_2}{\|b - Ax_0\|_2} = \frac{\|A(x^* - x_k)\|_2}{\|A(x^* - x_0)\|_2} \leq \sqrt{\frac{\lambda_1}{\lambda_N}}\frac{\|x^* - x_k\|_A}{\|x^* - x_0\|_A}$$

as asserted. □

So, to predict the performance of the CG iteration based on termination on small relative residuals, we must not only use (2.7) to predict when the relative

A-norm error is small, but also use Lemma 2.3.2 to relate small A-norm errors to small relative residuals.

We consider a very simple example. Assume that $x_0 = 0$ and that the eigenvalues of A are contained in the interval $(9, 11)$. If we let $\bar{p}_k(z) = (10 - z)^k/10^k$, then $\bar{p}_k \in \mathcal{P}_k$. This means that we may apply (2.7) to get

$$\|x_k - x^*\|_A \leq \|x^*\|_A \max_{9 \leq z \leq 11} |\bar{p}_k(z)|.$$

It is easy to see that

$$\max_{9 \leq z \leq 11} |\bar{p}_k(z)| = 10^{-k}.$$

Hence, after k iterations

$$(2.14) \qquad \|x_k - x^*\|_A \leq \|x^*\|_A 10^{-k}.$$

So, the size of the A-norm of the error will be reduced by a factor of 10^{-3} when

$$10^{-k} \leq 10^{-3},$$

that is, when

$$k \geq 3.$$

To use Lemma 2.3.2 we simply note that $\kappa_2(A) \leq 11/9$. Hence, after k iterations we have

$$\frac{\|Ax_k - b\|_2}{\|b\|_2} \leq \sqrt{11} \times 10^{-k}/3.$$

So, the size of the relative residual will be reduced by a factor of 10^{-3} when

$$10^{-k} \leq 3 \times 10^{-3}/\sqrt{11},$$

that is, when

$$k \geq 4.$$

One can obtain a more precise estimate by using a polynomial other than p_k in the upper estimate for the right-hand side of (2.7). Note that it is always the case that the spectrum of a spd matrix is contained in the interval $[\lambda_N, \lambda_1]$ and that $\kappa_2(A) = \lambda_1/\lambda_N$. A result from [48] (see also [45]) that is, in one sense, the sharpest possible, is

$$(2.15) \qquad \|x_k - x^*\|_A \leq 2\|x_0 - x^*\|_A \left[\frac{\sqrt{\kappa_2(A)} - 1}{\sqrt{\kappa_2(A)} + 1}\right]^k.$$

In the case of the above example, we can estimate $\kappa_2(A)$ by $\kappa_2(A) \leq 11/9$. Hence, since $(\sqrt{x} - 1)/(\sqrt{x} + 1)$ is an increasing function of x on the interval $(1, \infty)$.

$$\frac{\sqrt{\kappa_2(A)} - 1}{\sqrt{\kappa_2(A)} + 1} \leq \frac{\sqrt{11} - 3}{\sqrt{11} + 3} \approx .05.$$

Therefore (2.15) would predict a reduction in the size of the A-norm error by a factor of 10^{-3} when

$$2 \times .05^k < 10^{-3}$$

or when

$$k > -\log_{10}(2000)/\log_{10}(.05) \approx 3.3/1.3 \approx 2.6,$$

which also predicts termination within three iterations.

We may have more precise information than a single interval containing $\sigma(A)$. When we do, the estimate in (2.15) can be very pessimistic. If the eigenvalues *cluster* in a small number of intervals, the condition number can be quite large, but CG can perform very well. We will illustrate this with an example. Exercise 2.8.5 also covers this point.

Assume that $x_0 = 0$ and the eigenvalues of A lie in the two intervals $(1, 1.5)$ and $(399, 400)$. Based on this information the best estimate of the condition number of A is $\kappa_2(A) \leq 400$, which, when inserted into (2.15) gives

$$\frac{\|x_k - x^*\|_A}{\|x^*\|_A} \leq 2 \times (19/21)^k \approx 2 \times (.91)^k.$$

This would indicate fairly slow convergence. However, if we use as a residual polynomial $\bar{p}_{3k} \in \mathcal{P}_{3k}$

$$\bar{p}_{3k}(z) = \frac{(1.25 - z)^k (400 - z)^{2k}}{(1.25)^k \times 400^{2k}}.$$

It is easy to see that

$$\max_{z \in \sigma(A)} |\bar{p}_{3k}(z)| \leq (.25/1.25)^k = (.2)^k,$$

which is a sharper estimate on convergence. In fact, (2.15) would predict that

$$\|x_k - x^*\|_A \leq 10^{-3} \|x^*\|_A,$$

when $2 \times (.91)^k < 10^{-3}$ or when

$$k > -\log_{10}(2000)/\log_{10}(.91) \approx 3.3/.04 = 82.5.$$

The estimate based on the clustering gives convergence in $3k$ iterations when

$$(.2)^k \leq 10^{-3}$$

or when

$$k > -3/\log_{10}(.2) = 4.3.$$

Hence (2.15) predicts 83 iterations and the clustering analysis 15 (the smallest integer multiple of 3 larger than $3 \times 4.3 = 12.9$).

From the results above one can see that if the condition number of A is near one, the CG iteration will converge very rapidly. Even if the condition number is large, the iteration will perform well if the eigenvalues are clustered in a few small intervals. The transformation of the problem into one with eigenvalues clustered near one (i.e., easier to solve) is called *preconditioning*. We used this term before in the context of Richardson iteration and accomplished the goal by multiplying A by an approximate inverse. In the context of CG, such a simple approach can destroy the symmetry of the coefficient matrix and a more subtle implementation is required. We discuss this in § 2.5.

2.4. Implementation

The implementation of CG depends on the amazing fact that once x_k has been determined, either $x_k = x^*$ or a *search direction* $p_{k+1} \neq 0$ can be found very cheaply so that $x_{k+1} = x_k + \alpha_{k+1} p_{k+1}$ for some scalar α_{k+1}. Once p_{k+1} has been found, α_{k+1} is easy to compute from the minimization property of the iteration. In fact

(2.16) $$\frac{d\phi(x_k + \alpha p_{k+1})}{d\alpha} = 0$$

for the correct choice of $\alpha = \alpha_{k+1}$. Equation (2.16) can be written as

$$p_{k+1}^T A x_k + \alpha p_{k+1}^T A p_{k+1} - p_{k+1}^T b = 0$$

leading to

(2.17) $$\alpha_{k+1} = \frac{p_{k+1}^T (b - A x_k)}{p_{k+1}^T A p_{k+1}} = \frac{p_{k+1}^T r_k}{p_{k+1}^T A p_{k+1}}.$$

If $x_k = x_{k+1}$ then the above analysis implies that $\alpha = 0$. We show that this only happens if x_k is the solution.

LEMMA 2.4.1. *Let A be spd and let $\{x_k\}$ be the conjugate gradient iterates. Then*

(2.18) $$r_k^T r_l = 0 \text{ for all } 0 \leq l < k.$$

Proof. Since x_k minimizes ϕ on $x_0 + \mathcal{K}_k$, we have, for any $\xi \in \mathcal{K}_k$

$$\frac{d\phi(x_k + t\xi)}{dt} = \nabla\phi(x_k + t\xi)^T \xi = 0$$

at $t = 0$. Recalling that

$$\nabla\phi(x) = Ax - b = -r$$

we have

(2.19) $$\nabla\phi(x_k)^T \xi = -r_k^T \xi = 0 \text{ for all } \xi \in \mathcal{K}_k.$$

Since $r_l \in \mathcal{K}_k$ for all $l < k$ (see Exercise 2.8.1), this proves (2.18). $\square$

Now, if $x_k = x_{k+1}$, then $r_k = r_{k+1}$. Lemma 2.4.1 then implies that $\|r_k\|_2^2 = r_k^T r_k = r_k^T r_{k+1} = 0$ and hence $x_k = x^*$.

The next lemma characterizes the search direction and, as a side effect, proves that (if we define $p_0 = 0$) $p_l^T r_k = 0$ for all $0 \leq l < k \leq n$, unless the iteration terminates prematurely.

LEMMA 2.4.2. *Let A be spd and let $\{x_k\}$ be the conjugate gradient iterates. If $x_k \neq x^*$ then $x_{k+1} = x_k + \alpha_{k+1} p_{k+1}$ and p_{k+1} is determined up to a scalar multiple by the conditions*

$$(2.20) \qquad p_{k+1} \in \mathcal{K}_{k+1}, \; p_{k+1}^T A \xi = 0 \text{ for all } \xi \in \mathcal{K}_k.$$

Proof. Since $\mathcal{K}_k \subset \mathcal{K}_{k+1}$,

$$(2.21) \qquad \nabla \phi(x_{k+1})^T \xi = (A x_k + \alpha_{k+1} A p_{k+1} - b)^T \xi = 0$$

for all $\xi \in \mathcal{K}_k$. (2.19) and (2.21) then imply that for all $\xi \in \mathcal{K}_k$,

$$(2.22) \qquad \alpha_{k+1} p_{k+1}^T A \xi = -(A x_k - b)^T \xi = -\nabla \phi(x_k)^T \xi = 0.$$

This uniquely specifies the direction of p_{k+1} as (2.22) implies that $p_{k+1} \in \mathcal{K}_{k+1}$ is A-orthogonal (i.e., in the scalar product $(x, y) = x^T A y$) to $\mathcal{K}_k$, a subspace of dimension one less than $\mathcal{K}_{k+1}$. $\square$

The condition $p_{k+1}^T A \xi = 0$ is called *A-conjugacy* of p_{k+1} to $\mathcal{K}_k$. Now, any p_{k+1} satisfying (2.20) can, up to a scalar multiple, be expressed as

$$p_{k+1} = r_k + w_k$$

with $w_k \in \mathcal{K}_k$. While one might think that w_k would be hard to compute, it is, in fact, trivial. We have the following theorem.

THEOREM 2.4.1. *Let A be spd and assume that $r_k \neq 0$. Define $p_0 = 0$. Then*
$$(2.23) \qquad p_{k+1} = r_k + \beta_{k+1} p_k \text{ for some } \beta_{k+1} \text{ and } k \geq 0.$$

Proof. By Lemma 2.4.2 and the fact that $\mathcal{K}_k = \text{span}(r_0, \ldots, r_{k-1})$, we need only verify that a β_{k+1} can be found so that if p_{k+1} is given by (2.23) then

$$p_{k+1}^T A r_l = 0$$

for all $0 \leq l \leq k - 1$.

Let p_{k+1} be given by (2.23). Then for any $l \leq k$

$$p_{k+1}^T A r_l = r_k^T A r_l + \beta_{k+1} p_k^T A r_l.$$

If $l \leq k - 2$, then $r_l \in \mathcal{K}_{l+1} \subset \mathcal{K}_{k-1}$. Lemma 2.4.2 then implies that

$$p_{k+1}^T A r_l = 0 \text{ for } 0 \leq l \leq k - 2.$$

It only remains to solve for β_{k+1} so that $p_{k+1}^T A r_{k-1} = 0$. Trivially

$$(2.24) \qquad \beta_{k+1} = -r_k^T A r_{k-1} / p_k^T A r_{k-1}$$

provided $p_k^T A r_{k-1} \neq 0$. Since

$$r_k = r_{k-1} - \alpha_k A p_k$$

we have

$$r_k^T r_{k-1} = \|r_{k-1}\|_2^2 - \alpha_k p_k^T A r_{k-1}.$$

Since $r_k^T r_{k-1} = 0$ by Lemma 2.4.1 we have

$$(2.25) \qquad p_k^T A r_{k-1} = \|r_{k-1}\|_2^2 / \alpha_k \neq 0.$$

This completes the proof. $\square$

The common implementation of conjugate gradient uses a different form for α_k and β_k than given in (2.17) and (2.24).

LEMMA 2.4.3. *Let A be spd and assume that $r_k \neq 0$. Then*

$$(2.26) \qquad \alpha_{k+1} = \frac{\|r_k\|_2^2}{p_{k+1}^T A p_{k+1}}$$

and

$$(2.27) \qquad \beta_{k+1} = \frac{\|r_k\|_2^2}{\|r_{k-1}\|_2^2}.$$

Proof. Note that for $k \geq 0$

$$(2.28) \qquad p_{k+1}^T r_{k+1} = r_k^T r_{k+1} + \beta_{k+1} p_k^T r_{k+1} = 0$$

by Lemma 2.4.2. An immediate consequence of (2.28) is that $p_k^T r_k = 0$ and hence

$$(2.29) \qquad p_{k+1}^T r_k = (r_k + \beta_{k+1} p_k)^T r_k = \|r_k\|_2^2.$$

Taking scalar products of both sides of

$$r_{k+1} = r_k - \alpha_{k+1} A p_{k+1}$$

with p_{k+1} and using (2.29) gives

$$0 = p_{k+1}^T r_k - \alpha_{k+1} p_{k+1}^T A p_{k+1} = \|r_k^T\|_2^2 - \alpha_{k+1} p_{k+1}^T A p_{k+1},$$

which is equivalent to (2.26).

To get (2.27) note that $p_{k+1}^T A p_k = 0$ and hence (2.23) implies that

$$(2.30) \qquad \beta_{k+1} = \frac{-r_k^T A p_k}{p_k^T A p_k}.$$

Also note that

$$(2.31) \qquad \begin{aligned} p_k^T A p_k \ &= p_k^T A (r_{k-1} + \beta_k p_{k-1}) \\ &= p_k^T A r_{k-1} + \beta_k p_k^T A p_{k-1} = p_k^T A r_{k-1}. \end{aligned}$$

Now combine (2.30), (2.31), and (2.25) to get

$$\beta_{k+1} = \frac{-r_k^T A p_k \alpha_k}{\|r_{k-1}\|_2^2}.$$

Now take scalar products of both sides of

$$r_k = r_{k-1} - \alpha_k A p_k$$

with r_k and use Lemma 2.4.1 to get

$$\|r_k\|_2^2 = -\alpha_k r_k^T A p_k.$$

Hence (2.27) holds. □

The usual implementation reflects all of the above results. The goal is to find, for a given ϵ, a vector x so that $\|b - Ax\|_2 \le \epsilon \|b\|_2$. The input is the initial iterate x, which is overwritten with the solution, the right hand side b, and a routine which computes the action of A on a vector. We limit the number of iterations to $kmax$ and return the solution, which overwrites the initial iterate x and the residual norm.

ALGORITHM 2.4.1. $cg(x, b, A, \epsilon, kmax)$

1. $r = b - Ax$, $\rho_0 = \|r\|_2^2$, $k = 1$.

2. Do While $\sqrt{\rho_{k-1}} > \epsilon \|b\|_2$ and $k < kmax$

   (a) if $k = 1$ then $p = r$
       else
       $\beta = \rho_{k-1}/\rho_{k-2}$ and $p = r + \beta p$

   (b) $w = Ap$

   (c) $\alpha = \rho_{k-1}/p^T w$

   (d) $x = x + \alpha p$

   (e) $r = r - \alpha w$

   (f) $\rho_k = \|r\|_2^2$

   (g) $k = k + 1$

Note that the matrix A itself need not be formed or stored, only a routine for matrix-vector products is required. Krylov space methods are often called *matrix-free* for that reason.

Now, consider the costs. We need store only the four vectors x, w, p, and r. Each iteration requires a single matrix-vector product (to compute $w = Ap$), two scalar products (one for $p^T w$ and one to compute $\rho_k = \|r\|_2^2$), and three operations of the form $ax + y$, where x and y are vectors and a is a scalar.

It is remarkable that the iteration can progress without storing a basis for the entire Krylov subspace. As we will see in the section on GMRES, this is not the case in general. The spd structure buys quite a lot.

2.5. Preconditioning

To reduce the condition number, and hence improve the performance of the iteration, one might try to replace $Ax = b$ by another spd system with the same solution. If M is a spd matrix that is close to A^{-1}, then the eigenvalues of MA will be clustered near one. However MA is unlikely to be spd, and hence CG cannot be applied to the system $MAx = Mb$.

In theory one avoids this difficulty by expressing the preconditioned problem in terms of B, where B is spd, $A = B^2$, and by using a two-sided preconditioner, $S \approx B^{-1}$ (so $M = S^2$). Then the matrix SAS is spd and its eigenvalues are clustered near one. Moreover the preconditioned system

$$SASy = Sb$$

has $y^* = S^{-1}x^*$ as a solution, where $Ax^* = b$. Hence x^* can be recovered from y^* by multiplication by S. One might think, therefore, that computing S (or a subroutine for its action on a vector) would be necessary and that a matrix-vector multiply by SAS would incur a cost of one multiplication by A and two by S. Fortunately, this is not the case.

If y^k, $\hat{r}_k$, $\hat{p}_k$ are the iterate, residual, and search direction for CG applied to SAS and we let

$$x_k = S\hat{y}^k,\ r_k = S^{-1}\hat{r}_k,\ p_k = S\hat{p}_k,\ \text{and}\ z_k = S\hat{r}_k,$$

then one can perform the iteration directly in terms of x_k, A, and M. The reader should verify that the following algorithm does exactly that. The input is the same as that for Algorithm cg and the routine to compute the action of the preconditioner on a vector. Aside from the preconditioner, the arguments to pcg are the same as those to Algorithm cg.

ALGORITHM 2.5.1. pcg$(x, b, A, M, \epsilon, kmax)$

1. $r = b - Ax$, $\rho_0 = \|r\|_2^2$, $k = 1$

2. Do While $\sqrt{\rho_{k-1}} > \epsilon\|b\|_2$ and $k < kmax$

   (a) $z = Mr$

   (b) $\tau_{k-1} = z^T r$

   (c) if $k = 1$ then $\beta = 0$ and $p = z$
       else
       $\beta = \tau_{k-1}/\tau_{k-2},\ p = z + \beta p$

   (d) $w = Ap$

   (e) $\alpha = \tau_{k-1}/p^T w$

   (f) $x = x + \alpha p$

   (g) $r = r - \alpha w$

   (h) $\rho_k = r^T r$

   (i) $k = k + 1$

Note that the cost is identical to CG with the addition of

- the application of the preconditioner M in step 2a and

- the additional inner product required to compute τ_k in step 2b.

Of these costs, the application of the preconditioner is usually the larger. In the remainder of this section we briefly mention some classes of preconditioners. A more complete and detailed discussion of preconditioners is in [8] and a concise survey with many pointers to the literature is in [12].

Some effective preconditioners are based on deep insight into the structure of the problem. See [124] for an example in the context of partial differential equations, where it is shown that certain discretized second-order elliptic problems on simple geometries can be very well preconditioned with fast Poisson solvers [99], [188], and [187]. Similar performance can be obtained from multigrid [99], domain decomposition, [38], [39], [40], and alternating direction preconditioners [8], [149], [193], [194]. We use a Poisson solver preconditioner in the examples in § 2.7 and § 3.7 as well as for nonlinear problems in § 6.4.2 and § 8.4.2.

One commonly used and easily implemented preconditioner is Jacobi preconditioning, where M is the inverse of the diagonal part of A. One can also use other preconditioners based on the classical stationary iterative methods, such as the symmetric Gauss–Seidel preconditioner (1.18). For applications to partial differential equations, these preconditioners may be somewhat useful, but should not be expected to have dramatic effects.

Another approach is to apply a sparse Cholesky factorization to the matrix A (thereby giving up a fully matrix-free formulation) and discarding small elements of the factors and/or allowing only a fixed amount of storage for the factors. Such preconditioners are called *incomplete factorization* preconditioners. So if $A = LL^T + E$, where E is small, the preconditioner is $(LL^T)^{-1}$ and its action on a vector is done by two sparse triangular solves. We refer the reader to [8], [127], and [44] for more detail.

One could also attempt to estimate the spectrum of A, find a polynomial p such that $1 - zp(z)$ is small on the approximate spectrum, and use $p(A)$ as a preconditioner. This is called *polynomial preconditioning*. The preconditioned system is

$$p(A)Ax = p(A)b$$

and we would expect the spectrum of $p(A)A$ to be more clustered near $z = 1$ than that of A. If an interval containing the spectrum can be found, the residual polynomial $q(z) = 1 - zp(z)$ of smallest L^∞ norm on that interval can be expressed in terms of Chebyshev [161] polynomials. Alternatively q can be selected to solve a least squares minimization problem [5], [163]. The preconditioning p can be directly recovered from q and convergence rate estimates made. This technique is used to prove the estimate (2.15), for example. The cost of such a preconditioner, if a polynomial of degree K is used, is K matrix-vector products for each application of the preconditioner [5]. The performance gains can be very significant and the implementation is matrix-free.

2.6. CGNR and CGNE

If A is nonsingular and nonsymmetric, one might consider solving $Ax = b$ by applying CG to the normal equations

$$(2.32) \qquad\qquad A^T A x = A^T b.$$

This approach [103] is called CGNR [71], [78], [134]. The reason for this name is that the minimization property of CG as applied to (2.32) asserts that

$$
\begin{aligned}
\|x^* - x\|_{A^T A}^2 &= (x^* - x)^T A^T A (x^* - x) \\
&= (Ax^* - Ax)^T (Ax^* - Ax) = (b - Ax)^T (b - Ax) = \|r\|^2
\end{aligned}
$$

is minimized over $x_0 + \mathcal{K}_k$ at each iterate. Hence the name $\mathcal{C}$onjugate $\mathcal{G}$radient on the $\mathcal{N}$ormal equations to minimize the $\mathcal{R}$esidual.

Alternatively, one could solve

$$(2.33) \qquad\qquad AA^T y = b$$

and then set $x = A^T y$ to solve $Ax = b$. This approach [46] is now called CGNE [78], [134]. The reason for this name is that the minimization property of CG as applied to (2.33) asserts that if y^* is the solution to (2.33) then

$$
\begin{aligned}
\|y^* - y\|_{AA^T}^2 &= (y^* - y)^T (AA^T)(y^* - y) = (A^T y^* - A^T y)^T (A^T y^* - A^T y) \\[2mm]
&= \|x^* - x\|_2^2
\end{aligned}
$$

is minimized over $y_0 + \mathcal{K}_k$ at each iterate. $\mathcal{C}$onjugate $\mathcal{G}$radient on the $\mathcal{N}$ormal equations to minimize the $\mathcal{E}$rror.

The advantages of this approach are that all the theory for CG carries over and the simple implementation for both CG and PCG can be used. There are three disadvantages that may or may not be serious. The first is that the condition number of the coefficient matrix $A^T A$ is the square of that of A. The second is that two matrix-vector products are needed for each CG iterate since $w = A^T A p = A^T (Ap)$ in CGNR and $w = AA^T p = A(A^T p)$ in CGNE. The third, more important, disadvantage is that one must compute the action of A^T on a vector as part of the matrix-vector product involving $A^T A$. As we will see in the chapter on nonlinear problems, there are situations where this is not possible.

The analysis with residual polynomials is similar to that for CG. We consider the case for CGNR, the analysis for CGNE is essentially the same. As above, when we consider the $A^T A$ norm of the error, we have

$$\|x^* - x\|_{A^T A}^2 = (x^* - x)^T A^T A (x^* - x) = \|A(x^* - x)\|_2^2 = \|r\|_2^2.$$

Hence, for any residual polynomial $\bar{p}_k \in \mathcal{P}_k$,

$$(2.34) \qquad \|r_k\|_2 \le \|\bar{p}_k(A^T A) r_0\|_2 \le \|r_0\|_2 \max_{z \in \sigma(A^T A)} |\bar{p}_k(z)|.$$

There are two major differences between (2.34) and (2.7). The estimate is in terms of the l_2 norm of the residual, which corresponds exactly to the termination criterion, hence we need not prove a result like Lemma 2.3.2. Most significantly, the residual polynomial is to be maximized over the eigenvalues of $A^T A$, which is the set of the squares of the singular values of A. Hence the performance of CGNR and CGNE is determined by the distribution of singular values.

2.7. Examples for preconditioned conjugate iteration

In the collection of MATLAB codes we provide a code for preconditioned conjugate gradient iteration. The inputs, described in the comment lines, are the initial iterate, x_0, the right hand side vector b, MATLAB functions for the matrix-vector product and (optionally) the preconditioner, and iteration parameters to specify the maximum number of iterations and the termination criterion. On return the code supplies the approximate solution x and the history of the iteration as the vector of residual norms.

We consider the discretization of the partial differential equation

$$(2.35) \qquad\qquad -\nabla \cdot (a(x,y)\nabla u) = f(x,y)$$

on $0 < x, y < 1$ subject to homogeneous Dirichlet boundary conditions

$$u(x,0) = u(x,1) = u(0,y) = u(1,y) = 0, \quad 0 < x, y < 1.$$

One can verify [105] that the differential operator is positive definite in the Hilbert space sense and that the five-point discretization described below is positive definite if $a > 0$ for all $0 \le x, y \le 1$ (Exercise 2.8.10).

We discretize with a five-point centered difference scheme with n^2 points and mesh width $h = 1/(n+1)$. The unknowns are

$$u_{ij} \approx u(x_i, x_j)$$

where $x_i = ih$ for $1 \le i \le n$. We set

$$u_{0j} = u_{(n+1)j} = u_{i0} = u_{i(n+1)} = 0,$$

to reflect the boundary conditions, and define

$$\alpha_{ij} = -a(x_i, x_j)h^{-2}/2.$$

We express the discrete matrix-vector product as

$$(Au)_{ij} = (\alpha_{ij} + \alpha_{(i+1)j})(u_{(i+1)j} - u_{ij})$$

$$(2.36) \qquad -(\alpha_{(i-1)j} + \alpha_{ij})(u_{ij} - u_{(i-1)j}) + (\alpha_{i(j+1)} + \alpha_{ij})(u_{i(j+1)} - u_{ij})$$

$$-(\alpha_{ij} + \alpha_{i(j-1)})(u_{ij} - u_{i(j-1)})$$

for $1 \leq i, j \leq n$.

For the MATLAB implementation we convert freely from the representation of u as a two-dimensional array (with the boundary conditions added), which is useful for computing the action of A on u and applying fast solvers, and the representation as a one-dimensional array, which is what pcgsol expects to see. See the routine fish2d in collection of MATLAB codes for an example of how to do this in MATLAB.

For the computations reported in this section we took $a(x, y) = \cos(x)$ and took the right hand side so that the exact solution was the discretization of

$$10xy(1 - x)(1 - y) \exp(x^{4.5}).$$

The initial iterate was $u_0 = 0$.

In the results reported here we took $n = 31$ resulting in a system with $N = n^2 = 961$ unknowns. We expect second-order accuracy from the method and accordingly we set termination parameter $\epsilon = h^2 = 1/1024$. We allowed up to 100 CG iterations. The initial iterate was the zero vector. We will report our results graphically, plotting $\|r_k\|_2 / \|b\|_2$ on a semi-log scale.

In Fig. 2.1 the solid line is a plot of $\|r_k\| / \|b\|$ and the dashed line a plot of $\|u^* - u_k\|_A / \|u^* - u_0\|_A$. Note that the reduction in $\|r\|$ is not monotone. This is consistent with the theory, which predicts decrease in $\|e\|_A$ but not necessarily in $\|r\|$ as the iteration progresses. Note that the unpreconditioned iteration is slowly convergent. This can be explained by the fact that the eigenvalues are not clustered and

$$\kappa(A) = O(1/h^2) = O(n^2) = O(N)$$

and hence (2.15) indicates that convergence will be slow. The reader is asked to quantify this in terms of execution times in Exercise 2.8.9. This example illustrates the importance of a good preconditioner. Even the unpreconditioned iteration, however, is more efficient that the classical stationary iterative methods.

For a preconditioner we use a Poisson solver. By this we mean an operator G such that $v = Gw$ is the solution of the discrete form of

$$-v_{xx} - v_{yy} = w,$$

subject to homogeneous Dirichlet boundary conditions. The effectiveness of such a preconditioner has been analyzed in [124] and some of the many ways to implement the solver efficiently are discussed in [99], [188], [186], and [187].

The properties of CG on the preconditioned problem in the continuous case have been analyzed in [48]. For many types of domains and boundary conditions, Poisson solvers can be designed to take advantage of vector and/or parallel architectures or, in the case of the MATLAB environment used in this book, designed to take advantage of fast MATLAB built-in functions. Because of this their execution time is less than a simple count of floating-point operations would indicate. The fast Poisson solver in the collection of

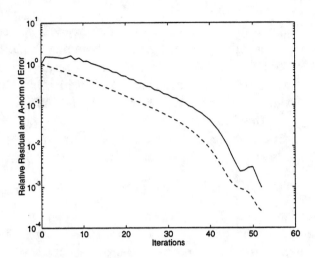

FIG. 2.1.   *CG for* 2-D *elliptic equation.*

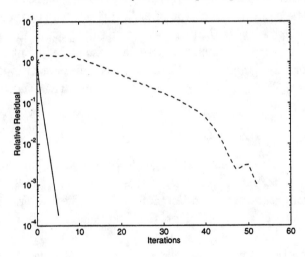

FIG. 2.2.   *PCG for* 2-D *elliptic equation.*

codes fish2d is based on the MATLAB fast Fourier transform, the built-in function fft.

In Fig. 2.2 the solid line is the graph of $\|r_k\|_2/\|b\|_2$ for the preconditioned iteration and the dashed line for the unpreconditioned. The preconditioned iteration required 5 iterations for convergence and the unpreconditioned iteration 52. Not only does the preconditioned iteration converge more rapidly, but the number of iterations required to reduce the relative residual by a given amount is independent of the mesh spacing [124]. We caution the reader that the preconditioned iteration is not as much faster than the

unpreconditioned one as the iteration count would suggest. The MATLAB
flops command indicates that the unpreconditioned iteration required roughly
1.2 million floating-point operations while the preconditioned iteration required
.87 million floating-point operations. Hence, the cost of the preconditioner is
considerable. In the MATLAB environment we used, the execution time of
the preconditioned iteration was about 60% of that of the unpreconditioned.
As we remarked above, this speed is a result of the efficiency of the MATLAB
fast Fourier transform. In Exercise 2.8.11 you are asked to compare execution
times for your own environment.

2.8. Exercises on conjugate gradient

2.8.1. Let $\{x_k\}$ be the conjugate gradient iterates. Prove that $r_l \in \mathcal{K}_k$ for all $l < k$.

2.8.2. Let A be spd. Show that there is a spd B such that $B^2 = A$. Is B unique?

2.8.3. Let Λ be a diagonal matrix with $\Lambda_{ii} = \lambda_i$ and let p be a polynomial. Prove that $\|p(\Lambda)\| = \max_i |p(\lambda_i)|$ where $\|\cdot\|$ is any induced matrix norm.

2.8.4. Prove Theorem 2.2.3.

2.8.5. Assume that A is spd and that

$$\sigma(A) \subset (1, 1.1) \cup (2, 2.2).$$

Give upper estimates based on (2.6) for the number of CG iterations required to reduce the A norm of the error by a factor of 10^{-3} and for the number of CG iterations required to reduce the residual by a factor of 10^{-3}.

2.8.6. For the matrix A in problem 5, assume that the cost of a matrix vector multiply is $4N$ floating-point multiplies. Estimate the number of floating-point operations reduce the A norm of the error by a factor of 10^{-3} using CG iteration.

2.8.7. Let A be a nonsingular matrix with all singular values in the interval $(1, 2)$. Estimate the number of CGNR/CGNE iterations required to reduce the relative residual by a factor of 10^{-4}.

2.8.8. Show that if A has constant diagonal then PCG with Jacobi preconditioning produces the same iterates as CG with no preconditioning.

2.8.9. Assume that A is $N \times N$, nonsingular, and spd. If $\kappa(A) = O(N)$, give a rough estimate of the number of CG iterates required to reduce the relative residual to $O(1/N)$.

2.8.10. Prove that the linear transformation given by (2.36) is symmetric and positive definite on R^{n^2} if $a(x, y) > 0$ for all $0 \leq x, y \leq 1$.

2.8.11. Duplicate the results in § 2.7 for example, in MATLAB by writing the matrix-vector product routines and using the MATLAB codes pcgsol and fish2d. What happens as N is increased? How are the performance and accuracy affected by changes in $a(x, y)$? Try $a(x, y) = \sqrt{.1 + x}$ and examine the accuracy of the result. Explain your findings. Compare the execution times on your computing environment (using the cputime command in MATLAB, for instance).

2.8.12. Use the Jacobi and symmetric Gauss–Seidel iterations from Chapter 1 to solve the elliptic boundary value problem in § 2.7. How does the performance compare to CG and PCG?

2.8.13. Implement Jacobi (1.17) and symmetric Gauss–Seidel (1.18) preconditioners for the elliptic boundary value problem in § 2.7. Compare the performance with respect to both computer time and number of iterations to preconditioning with the Poisson solver.

2.8.14. Modify `pcgsol` so that $\phi(x)$ is computed and stored at each iterate and returned on output. Plot $\phi(x_n)$ as a function of n for each of the examples.

2.8.15. Apply CG and PCG to solve the five-point discretization of

$$-u_{xx}(x,y) - u_{yy}(x,y) + e^{x+y}u(x,y) = 1,\, 0 < x, y, < 1,$$

subject to the *inhomogeneous* Dirichlet boundary conditions

$$u(x,0) = u(x,1) = u(1,y) = 0, u(0,y) = 1, \quad 0 < x, y < 1.$$

Experiment with different mesh sizes and preconditioners (Fast Poisson solver, Jacobi, and symmetric Gauss–Seidel).

Chapter 3

GMRES Iteration

3.1. The minimization property and its consequences

The GMRES (Generalized Minimum RESidual) was proposed in 1986 in [167] as a Krylov subspace method for nonsymmetric systems. Unlike CGNR, GMRES does not require computation of the action of A^T on a vector. This is a significant advantage in many cases. The use of residual polynomials is made more complicated because we cannot use the spectral theorem to decompose A. Moreover, one must store a basis for $\mathcal{K}_k$, and therefore storage requirements increase as the iteration progresses.

The kth ($k \geq 1$) iteration of GMRES is the solution to the least squares problem

(3.1) $$\text{minimize}_{x \in x_0 + \mathcal{K}_k} \|b - Ax\|_2.$$

The beginning of this section is much like the analysis for CG. Note that if $x \in x_0 + \mathcal{K}_k$ then

$$x = x_0 + \sum_{j=0}^{k-1} \gamma_j A^j r_0$$

and so

$$b - Ax = b - Ax_0 - \sum_{j=0}^{k-1} \gamma_j A^{j+1} r_0 = r_0 - \sum_{j=1}^{k} \gamma_{j-1} A^j r_0.$$

Hence if $x \in x_0 + \mathcal{K}_k$ then $r = \bar{p}(A) r_0$ where $\bar{p} \in \mathcal{P}_k$ is a residual polynomial. We have just proved the following result.

THEOREM 3.1.1. *Let A be nonsingular and let x_k be the kth GMRES iteration. Then for all $\bar{p}_k \in \mathcal{P}_k$*

(3.2) $$\|r_k\|_2 = \min_{p \in \mathcal{P}_k} \|\bar{p}(A) r_0\|_2 \leq \|\bar{p}_k(A) r_0\|_2.$$

From this we have the following corollary.

COROLLARY 3.1.1. *Let A be nonsingular and let x_k be the kth GMRES iteration. Then for all $\bar{p}_k \in \mathcal{P}_k$*

(3.3) $$\frac{\|r_k\|_2}{\|r_0\|_2} \leq \|\bar{p}_k(A)\|_2.$$

We can apply the corollary to prove finite termination of the GMRES iteration.

THEOREM 3.1.2. *Let A be nonsingular. Then the GMRES algorithm will find the solution within N iterations.*

Proof. The *characteristic polynomial* of A is $p(z) = \det(A - zI)$. p has degree N, $p(0) = det(A) \neq 0$ since A is nonsingular, and so

$$\bar{p}_N(z) = p(z)/p(0) \in \mathcal{P}_N$$

is a residual polynomial. It is well known [141] that $p(A) = \bar{p}_N(A) = 0$. By (3.3), $r_N = b - Ax_N = 0$ and hence x_N is the solution. □

In Chapter 2 we applied the spectral theorem to obtain more precise information on convergence rates. This is not an option for general nonsymmetric matrices. However, if A is diagonalizable we may use (3.2) to get information from clustering of the spectrum just like we did for CG. We pay a price in that we must use complex arithmetic for the only time in this book. Recall that A is *diagonalizable* if there is a nonsingular (*possibly complex!*) matrix V such that

$$A = V\Lambda V^{-1}.$$

Here Λ is a (*possibly complex!*) diagonal matrix with the eigenvalues of A on the diagonal. If A is a diagonalizable matrix and p is a polynomial then

$$p(A) = Vp(\Lambda)V^{-1}$$

A is *normal* if the *diagonalizing transformation* V is orthogonal. In that case the columns of V are the eigenvectors of A and $V^{-1} = V^H$. Here V^H is the complex conjugate transpose of V. In the remainder of this section we must use complex arithmetic to analyze the convergence. Hence we will switch to complex matrices and vectors. Recall that the scalar product in C^N, the space of complex N-vectors, is $x^H y$. In particular, we will use the l^2 norm in C^N. Our use of complex arithmetic will be implicit for the most part and is needed only so that we may admit the possibility of complex eigenvalues of A.

We can use the structure of a diagonalizable matrix to prove the following result.

THEOREM 3.1.3. *Let $A = V\Lambda V^{-1}$ be a nonsingular diagonalizable matrix. Let x_k be the kth GMRES iterate. Then for all $\bar{p}_k \in \mathcal{P}_k$*

(3.4) $$\frac{\|r_k\|_2}{\|r_0\|_2} \leq \kappa_2(V) \max_{z \in \sigma(A)} |\bar{p}_k(z)|.$$

Proof. Let $\bar{p}_k \in \mathcal{P}_k$. We can easily estimate $\|\bar{p}_k(A)\|_2$ by

$$\|\bar{p}_k(A)\|_2 \leq \|V\|_2 \|V^{-1}\|_2 \|\bar{p}_k(\Lambda)\|_2 \leq \kappa_2(V) \max_{z \in \sigma(A)} |\bar{p}_k(z)|,$$

as asserted. □

It is not clear how one should estimate the condition number of the diagonalizing transformation if it exists. If A is normal, of course, $\kappa_2(V) = 1$.

As we did for CG, we look at some easy consequences of (3.3) and (3.4).

THEOREM 3.1.4. *Let A be a nonsingular diagonalizable matrix. Assume that A has only k distinct eigenvalues. Then GMRES will terminate in at most k iterations.*

THEOREM 3.1.5. *Let A be a nonsingular normal matrix. Let b be a linear combination of k of the eigenvectors of A*

$$b = \sum_{l=1}^{k} \gamma_l u_{i_l}.$$

Then the GMRES iteration for $Ax = b$ will terminate in at most k iterations.

3.2. Termination

As is the case with CG, GMRES is best thought of as an iterative method. The convergence rate estimates for the diagonalizable case will involve $\kappa_2(V)$, but will otherwise resemble those for CG. If A is not diagonalizable, rate estimates have been derived in [139], [134], [192], [33], and [34]. As the set of nondiagonalizable matrices has measure zero in the space of $N \times N$ matrices, the chances are very high that a computed matrix will be diagonalizable. This is particularly so for the finite difference Jacobian matrices we consider in Chapters 6 and 8. Hence we confine our attention to diagonalizable matrices.

As was the case with CG, we terminate the iteration when

$$(3.5) \qquad \qquad \|r_k\|_2 \leq \eta \|b\|_2$$

for the purposes of this example. We can use (3.3) and (3.4) directly to estimate the first k such that (3.5) holds without requiring a lemma like Lemma 2.3.2.

Again we look at examples. Assume that $A = V\Lambda V^{-1}$ is diagonalizable, that the eigenvalues of A lie in the interval $(9, 11)$, and that $\kappa_2(V) = 100$. We assume that $x_0 = 0$ and hence $r_0 = b$. Using the residual polynomial $\bar{p}_k(z) = (10 - z)^k / 10^k$ we find

$$\frac{\|r_k\|_2}{\|r_0\|_2} \leq (100)10^{-k} = 10^{2-k}.$$

Hence (3.5) holds when $10^{2-k} < \eta$ or when

$$k > 2 + \log_{10}(\eta).$$

Assume that $\|I - A\|_2 \leq \rho < 1$. Let $\bar{p}_k(z) = (1 - z)^k$. It is a direct consequence of (3.2) that

$$(3.6) \qquad \qquad \|r_k\|_2 \leq \rho^k \|r_0\|_2.$$

The estimate (3.6) illustrates the potential benefits of a good approximate inverse preconditioner.

The convergence estimates for GMRES in the nonnormal case are much less satisfying that those for CG, CGNR, CGNE, or GMRES in the normal case. This is a very active area of research and we refer to [134], [33], [120], [34], and [36] for discussions of and pointers to additional references to several questions related to nonnormal matrices.

3.3. Preconditioning

Preconditioning for GMRES and other methods for nonsymmetric problems is different from that for CG. There is no concern that the preconditioned system be spd and hence (3.6) essentially tells the whole story. However there are two different ways to view preconditioning. If one can find M such that

$$\|I - MA\|_2 = \rho < 1,$$

then applying GMRES to $MAx = Mb$ allows one to apply (3.6) to the preconditioned system.   Preconditioning done in this way is called *left preconditioning*. If $r = MAx - Mb$ is the residual for the preconditioned system, we have, if the product MA can be formed without error,

$$\frac{\|e_k\|_2}{\|e_0\|_2} \leq \kappa_2(MA)\frac{\|r_k\|_2}{\|r_0\|_2},$$

by Lemma 1.1.1. Hence, if MA has a smaller condition number than A, we might expect the relative residual of the preconditioned system to be a better indicator of the relative error than the relative residual of the original system. If

$$\|I - AM\|_2 = \rho < 1,$$

one can solve the system $AMy = b$ with GMRES and then set $x = My$. This is called *right preconditioning*. The residual of the preconditioned problem is the same as that of the unpreconditioned problem. Hence, the value of the relative residuals as estimators of the relative error is unchanged. Right preconditioning has been used as the basis for a method that changes the preconditioner as the iteration progresses [166].

One important aspect of implementation is that, unlike PCG, one can apply the algorithm directly to the system $MAx = Mb$ (or $AMy = b$). Hence, one can write a single matrix-vector product routine for MA (or AM) that includes both the application of A to a vector and that of the preconditioner.

Most of the preconditioning ideas mentioned in § 2.5 are useful for GMRES as well. In the examples in § 3.7 we use the Poisson solver preconditioner for nonsymmetric partial differential equations. Multigrid [99] and alternating direction [8], [182] methods have similar performance and may be more generally applicable. Incomplete factorization (LU in this case) preconditioners can be used [165] as can polynomial preconditioners. Some hybrid algorithms use the GMRES/Arnoldi process itself to construct polynomial preconditioners for GMRES or for Richardson iteration [135], [72], [164], [183]. Again we mention [8] and [12] as a good general references for preconditioning.

3.4. GMRES implementation: Basic ideas

Recall that the least squares problem defining the kth GMRES iterate is

$$\text{minimize}_{x \in x_0 + \mathcal{K}_k} \|b - Ax\|_2.$$

Suppose one had an orthogonal projector V_k onto $\mathcal{K}_k$. Then any $z \in \mathcal{K}_k$ can be written as

$$z = \sum_{l=1}^{k} y_l v_l^k,$$

where v_l^k is the lth column of V_k. Hence we can convert (3.1) to a least squares problem in R^k for the coefficient vector y of $z = x - x_0$. Since

$$x - x_0 = V_k y$$

for some $y \in R^k$, we must have $x_k = x_0 + V_k y$ where y minimizes

$$\|b - A(x_0 + V_k y)\|_2 = \|r_0 - AV_k y\|_2.$$

Hence, our least squares problem in R^k is

$$(3.7) \qquad\qquad \text{minimize}_{y \in R^k} \|r_0 - AV_k y\|_2.$$

This is a standard linear least squares problem that could be solved by a QR factorization, say. The problem with such a direct approach is that the matrix vector product of A with the basis matrix V_k must be taken at each iteration.

   If one uses Gram–Schmidt orthogonalization, however, one can represent (3.7) very efficiently and the resulting least squares problem requires no extra multiplications of A with vectors. The Gram–Schmidt procedure for formation of an orthonormal basis for $\mathcal{K}_k$ is called the Arnoldi [4] process. The data are vectors x_0 and b, a map that computes the action of A on a vector, and a dimension k. The algorithm computes an orthonormal basis for $\mathcal{K}_k$ and stores it in the columns of V.

   ALGORITHM 3.4.1. $\texttt{arnoldi}(x_0, b, A, k, V)$
   1. Define $r_0 = b - Ax_0$ and $v_1 = r_0/\|r_0\|_2$.

   2. For $i = 1, \ldots, k - 1$

$$v_{i+1} = \frac{Av_i - \sum_{j=1}^{i}((Av_i)^T v_j)v_j}{\|Av_i - \sum_{j=1}^{i}((Av_i)^T v_j)v_j\|_2}$$

   If there is never a division by zero in step 2 of Algorithm $\texttt{arnoldi}$, then the columns of the matrix V_k are an orthonormal basis for $\mathcal{K}_k$. A division by zero is referred to as *breakdown* and happens only if the solution to $Ax = b$ is in $\mathcal{K}_{k-1}$.

LEMMA 3.4.1. *Let A be nonsingular, let the vectors v_j be generated by Algorithm* arnoldi, *and let i be the smallest integer for which*

$$Av_i - \sum_{j=1}^{i}((Av_i)^T v_j)v_j = 0.$$

Then $x = A^{-1}b \in x_0 + \mathcal{K}_i$.

Proof. By hypothesis $Av_i \in \mathcal{K}_i$ and hence $A\mathcal{K}_i \subset \mathcal{K}_i$. Since the columns of V_i are an orthonormal basis for $\mathcal{K}_i$, we have

$$AV_i = V_i H,$$

where H is an $i \times i$ matrix. H is nonsingular since A is. Setting $\beta = \|r_0\|_2$ and $e_1 = (1, 0, \ldots, 0)^T \in R^i$ we have

$$\|r_i\|_2 = \|b - Ax_i\|_2 = \|r_0 - A(x_i - x_0)\|_2.$$

Since $x_i - x_0 \in \mathcal{K}_i$ there is $y \in R^i$ such that $x_i - x_0 = V_i y$. Since $r_0 = \beta V_i e_1$ and V_i is an orthogonal matrix

$$\|r_i\|_2 = \|V_i(\beta e_1 - Hy)\|_2 = \|\beta e_1 - Hy\|_{R^{i+1}},$$

where $\| \cdot \|_{R^{k+1}}$ denotes the Euclidean norm in R^{k+1}.

Setting $y = \beta H^{-1}e_1$ proves that $r_i = 0$ by the minimization property. $\square$

The upper Hessenberg structure can be exploited to make the solution of the least squares problems very efficient [167].

If the Arnoldi process does not break down, we can use it to implement GMRES in an efficient way. Set $h_{ji} = (Av_j)^T v_i$. By the Gram–Schmidt construction, the $k+1 \times k$ matrix H_k whose entries are h_{ij} is *upper Hessenberg*, i.e., $h_{ij} = 0$ if $i > j+1$. The Arnoldi process (unless it terminates prematurely with a solution) produces matrices $\{V_k\}$ with orthonormal columns such that

$$(3.8) \qquad AV_k = V_{k+1}H_k.$$

Hence, for some $y^k \in R^k$,

$$r_k = b - Ax_k = r_0 - A(x_k - x_0) = V_{k+1}(\beta e_1 - H_k y^k).$$

Hence $x_k = x_0 + V_k y^k$, where y^k minimizes $\|\beta e_1 - H_k y\|_2$ over R^k. Note that when y^k has been computed, the norm of r_k can be found without explicitly forming x_k and computing $r_k = b - Ax_k$. We have, using the orthogonality of V_{k+1},

$$(3.9) \qquad \|r_k\|_2 = \|V_{k+1}(\beta e_1 - H_k y^k)\|_2 = \|\beta e_1 - H_k y^k\|_{R^{k+1}}.$$

The goal of the iteration is to find, for a given ϵ, a vector x so that

$$\|b - Ax\|_2 \leq \epsilon \|b\|_2.$$

The input is the initial iterate, x, the right-hand side b, and a map that computes the action of A on a vector. We limit the number of iterations to *kmax* and return the solution, which overwrites the initial iterate x and the residual norm.

ALGORITHM 3.4.2. $\text{gmresa}(x, b, A, \epsilon, kmax, \rho)$
1. $r = b - Ax$, $v_1 = r/\|r\|_2$, $\rho = \|r\|_2$, $\beta = \rho$, $k = 0$

2. While $\rho > \epsilon\|b\|_2$ and $k < kmax$ do

   (a) $k = k + 1$

   (b) for $j = 1, \ldots, k$
   $$h_{jk} = (Av_k)^T v_j$$

   (c) $v_{k+1} = Av_k - \sum_{j=1}^{k} h_{jk}v_j$

   (d) $h_{k+1,k} = \|v_{k+1}\|_2$

   (e) $v_{k+1} = v_{k+1}/\|v_{k+1}\|_2$

   (f) $e_1 = (1, 0, \ldots, 0)^T \in R^{k+1}$
   Minimize $\|\beta e_1 - H_k y^k\|_{R^{k+1}}$ over R^k to obtain y^k.

   (g) $\rho = \|\beta e_1 - H_k y^k\|_{R^{k+1}}$.

3. $x_k = x_0 + V_k y^k$.

Note that x_k is only computed upon termination and is not needed within the iteration. It is an important property of GMRES that the basis for the Krylov space must be stored as the iteration progress. This means that in order to perform k GMRES iterations one must store k vectors of length N. For very large problems this becomes prohibitive and the iteration is *restarted* when the available room for basis vectors is exhausted. One way to implement this is to set $kmax$ to the maximum number m of vectors that one can store, call GMRES and explicitly test the residual $b - Ax_k$ if $k = m$ upon termination. If the norm of the residual is larger than ϵ, call GMRES again with $x_0 = x_k$, the result from the previous call. This restarted version of the algorithm is called GMRES(m) in [167]. There is no general convergence theorem for the restarted algorithm and restarting will slow the convergence down. However, when it works it can significantly reduce the storage costs of the iteration. We discuss implementation of GMRES(m) later in this section.

Algorithm gmresa can be implemented very straightforwardly in MATLAB. Step 2f can be done with a single MATLAB command, the backward division operator, at a cost of $O(k^3)$ floating-point operations. There are more efficient ways to solve the least squares problem in step 2f, [167], [197], and we use the method of [167] in the collection of MATLAB codes. The savings are slight if k is small relative to N, which is often the case for large problems, and the simple one-line MATLAB approach can be efficient for such problems.

A more serious problem with the implementation proposed in Algorithm gmresa is that the vectors v_j may become nonorthogonal as a result of cancellation errors. If this happens, (3.9), which depends on this orthogonality, will not hold and the residual and approximate solution could be inaccurate. A partial remedy is to replace the classical Gram–Schmidt orthogonalization in Algorithm gmresa with *modified* Gram–Schmidt orthogonalization. We replace the loop in step 2c of Algorithm gmresa with

$$v_{k+1} = Av_k$$
$$\text{for } j = 1, \ldots k$$
$$v_{k+1} = v_{k+1} - (v_{k+1}^T v_j)v_j.$$

While modified Gram–Schmidt and classical Gram–Schmidt are equivalent in infinite precision, the modified form is much more likely in practice to maintain orthogonality of the basis.

We illustrate this point with a simple example from [128], doing the computations in MATLAB. Let $\delta = 10^{-7}$ and define

$$A = \begin{pmatrix} 1 & 1 & 1 \\ \delta & \delta & 0 \\ \delta & 0 & \delta \end{pmatrix}.$$

We orthogonalize the columns of A with classical Gram–Schmidt to obtain

$$V = \begin{pmatrix} 1.0000e+00 & 1.0436e-07 & 9.9715e-08 \\ 1.0000e-07 & 1.0456e-14 & -9.9905e-01 \\ 1.0000e-07 & -1.0000e+00 & 4.3568e-02 \end{pmatrix}.$$

The columns of V_U are not orthogonal at all. In fact $v_2^T v_3 \approx -.004$. For modified Gram–Schmidt

$$V = \begin{pmatrix} 1.0000e+00 & 1.0436e-07 & 1.0436e-07 \\ 1.0000e-07 & 1.0456e-14 & -1.0000e+00 \\ 1.0000e-07 & -1.0000e+00 & 4.3565e-16 \end{pmatrix}.$$

Here $|v_i^T v_j - \delta_{ij}| \leq 10^{-8}$ for all i, j.

The versions we implement in the collection of MATLAB codes use modified Gram–Schmidt. The outline of our implementation is Algorithm gmresb. This implementation solves the upper Hessenberg least squares problem using the MATLAB backward division operator, and is not particularly efficient. We present a better implementation in Algorithm gmres. However, this version is very simple and illustrates some important ideas. First, we see that x_k need only be computed after termination as the least squares residual ρ can be used to approximate the norm of the residual (they are identical in exact arithmetic). Second, there is an opportunity to compensate for a loss of orthogonality in the basis vectors for the Krylov space. One can take a second pass through the modified Gram–Schmidt process and restore lost orthogonality [147], [160].

ALGORITHM 3.4.3. gmresb$(x, b, A, \epsilon, kmax, \rho)$
1. $r = b - Ax$, $v_1 = r/\|r\|_2$, $\rho = \|r\|_2$, $\beta = \rho$, $k = 0$

2. While $\rho > \epsilon\|b\|_2$ and $k < kmax$ do

   (a) $k = k + 1$

(b) $v_{k+1} = Av_k$
for $j = 1, \ldots k$
   i. $h_{jk} = v_{k+1}^T v_j$
   ii. $v_{k+1} = v_{k+1} - h_{jk}v_j$

(c) $h_{k+1,k} = \|v_{k+1}\|_2$

(d) $v_{k+1} = v_{k+1}/\|v_{k+1}\|_2$

(e) $e_1 = (1, 0, \ldots, 0)^T \in R^{k+1}$
Minimize $\|\beta e_1 - H_k y^k\|_{R^{k+1}}$ to obtain $y^k \in R^k$.

(f) $\rho = \|\beta e_1 - H_k y^k\|_{R^{k+1}}$.

3. $x_k = x_0 + V_k y^k$.

Even if modified Gram–Schmidt orthogonalization is used, one can still lose orthogonality in the columns of V. One can test for loss of orthogonality [22], [147], and reorthogonalize if needed or use a more stable means to create the matrix V [195]. These more complex implementations are necessary if A is ill conditioned or many iterations will be taken. For example, one can augment the modified Gram–Schmidt process

- $v_{k+1} = Av_k$
  for $j = 1, \ldots k$
  $h_{jk} = v_{k+1}^T v_j$
  $v_{k+1} = v_{k+1} - h_{jk}v_j$

- $h_{k+1,k} = \|v_{k+1}\|_2$

- $v_{k+1} = v_{k+1}/\|v_{k+1}\|_2$

with a second pass (reorthogonalization). One can reorthogonalize in every iteration or only if a test [147] detects a loss of orthogonality. There is nothing to be gained by reorthogonalizing more than once [147].

The modified Gram–Schmidt process with reorthogonalization looks like

- $v_{k+1} = Av_k$
  for $j = 1, \ldots, k$
  $h_{jk} = v_{k+1}^T v_j$
  $v_{k+1} = v_{k+1} - h_{jk}v_j$

- $h_{k+1,k} = \|v_{k+1}\|_2$

- If loss of orthogonality is detected
  For $j = 1, \ldots, k$
  $h_{tmp} = v_{k+1}^T v_j$
  $h_{jk} = h_{jk} + h_{tmp}$
  $v_{k+1} = v_{k+1} - h_{tmp}v_j$

- $h_{k+1,k} = \|v_{k+1}\|_2$

- $v_{k+1} = v_{k+1}/\|v_{k+1}\|_2$

One approach to reorthogonalization is to reorthogonalize in every step. This doubles the cost of the computation of V and is usually unnecessary. More efficient and equally effective approaches are based on other ideas. A variation on a method from [147] is used in [22]. Reorthogonalization is done after the Gram–Schmidt loop and before v_{k+1} is normalized if

$$(3.10) \qquad \|Av_k\|_2 + \delta\|v_{k+1}\|_2 = \|Av_k\|_2$$

to working precision. The idea is that if the new vector is very small relative to Av_k then information may have been lost and a second pass through the modified Gram–Schmidt process is needed. We employ this test in the MATLAB code gmres with $\delta = 10^{-3}$.

To illustrate the effects of loss of orthogonality and those of reorthogonalization we apply GMRES to the diagonal system $Ax = b$ where $b = (1,1,1)^T$, $x_0 = (0,0,0)^T$, and

$$(3.11) \qquad A = \begin{pmatrix} .001 & 0 & 0 \\ 0 & .0011 & 0 \\ 0 & 0 & 10^4 \end{pmatrix}.$$

While in infinite precision arithmetic only three iterations are needed to solve the system exactly, we find in the MATLAB environment that a solution to full precision requires more than three iterations unless reorthogonalization is applied after every iteration. In Table 3.1 we tabulate relative residuals as a function of the iteration counter for classical Gram–Schmidt without reorthogonalization (CGM), modified Gram–Schmidt without reorthogonalization (MGM), reorthogonalization based on the test (3.10) (MGM-PO), and reorthogonalization in every iteration (MGM-FO). While classical Gram–Schmidt fails, the reorthogonalization strategy based on (3.10) is almost as effective as the much more expensive approach of reorthogonalizing in every step. The method based on (3.10) is the default in the MATLAB code gmresa.

The kth GMRES iteration requires a matrix-vector product, k scalar products, and the solution of the Hessenberg least squares problem in step 2e. The k scalar products require $O(kN)$ floating-point operations and the cost of a solution of the Hessenberg least squares problem, by QR factorization or the MATLAB backward division operator, say, in step 2e of gmresb is $O(k^3)$ floating-point operations. Hence the total cost of the m GMRES iterations is m matrix-vector products and $O(m^4 + m^2N)$ floating-point operations. When k is not too large and the cost of matrix-vector products is high, a brute-force solution to the least squares problem using the MATLAB backward division operator is not terribly inefficient. We provide an implementation of Algorithm gmresb in the collection of MATLAB codes. This is an appealing algorithm, especially when implemented in an environment like MATLAB, because of its simplicity. For large k, however, the brute-force method can be very costly.

TABLE 3.1
Effects of reorthogonalization.

k	CGM	MGM	MGM-PO	MGM-FO
0	1.00e+00	1.00e+00	1.00e+00	1.00e+00
1	8.16e-01	8.16e-01	8.16e-01	8.16e-01
2	3.88e-02	3.88e-02	3.88e-02	3.88e-02
3	6.69e-05	6.42e-08	6.42e-08	6.34e-34
4	4.74e-05	3.70e-08	5.04e-24	
5	3.87e-05	3.04e-18		
6	3.35e-05			
7	3.00e-05			
8	2.74e-05			
9	2.53e-05			
10	2.37e-05			

3.5. Implementation: Givens rotations

If k is large, implementations using Givens rotations [167], [22], Householder reflections [195], or a shifted Arnoldi process [197] are much more efficient than the brute-force approach in Algorithm gmresb. The implementation in Algorithm gmres and in the MATLAB code collection is from [167]. This implementation maintains the QR factorization of H_k in a clever way so that the cost for a single GMRES iteration is $O(Nk)$ floating-point operations. The $O(k^2)$ cost of the triangular solve and the $O(kN)$ cost of the construction of x_k are incurred after termination.

A 2×2 *Givens rotation* is a matrix of the form

$$(3.12) \qquad G = \begin{pmatrix} c & -s \\ s & c \end{pmatrix},$$

where $c = \cos(\theta)$, $s = \sin(\theta)$ for $\theta \in [-\pi, \pi]$. The orthogonal matrix G rotates the vector $(c, -s)$, which makes an angle of $-\theta$ with the x-axis through an angle θ so that it overlaps the x-axis.

$$G \begin{pmatrix} c \\ -s \end{pmatrix} = \begin{pmatrix} 1 \\ 0 \end{pmatrix}.$$

An $N \times N$ Givens rotation replaces a 2×2 block on the diagonal of the

$N \times N$ identity matrix with a 2×2 Givens rotation.

$$
(3.13) \qquad G = \begin{pmatrix}
1 & 0 & & \cdots & & & 0 \\
0 & \ddots & \ddots & & & & \\
& & \ddots & c & -s & & \\
\vdots & & & s & c & 0 & \vdots \\
& & & & 0 & 1 & \ddots \\
& & & & & \ddots & \ddots & 0 \\
0 & & & \cdots & & 0 & 1
\end{pmatrix}.
$$

Our notation is that $G_j(c, s)$ is an $N \times N$ givens rotation of the form (3.13) with a 2×2 Givens rotation in rows and columns j and $j + 1$.

Givens rotations are used to annihilate single nonzero elements of matrices in reduction to triangular form [89]. They are of particular value in reducing Hessenberg matrices to triangular form and thereby solving Hessenberg least squares problems such as the ones that arise in GMRES. This reduction can be accomplished in $O(N)$ floating-point operations and hence is far more efficient than a solution by a singular value decomposition or a reduction based on Householder transformations. This method is also used in the QR algorithm for computing eigenvalues [89], [184].

Let H be an $N \times M$ $(N \geq M)$ upper Hessenberg matrix with rank M. We reduce H to triangular form by first multiplying the matrix by a Givens rotation that annihilates h_{21} (and, of course, changes h_{11} and the subsequent columns). We define $G_1 = G_1(c_1, s_1)$ by

$$
(3.14) \qquad c_1 = h_{11}/\sqrt{h_{11}^2 + h_{21}^2} \quad \text{and} \quad s_1 = -h_{21}/\sqrt{h_{11}^2 + h_{21}^2}.
$$

If we replace H by $G_1 H$, then the first column of H now has only a single nonzero element h_{11}. Similarly, we can now apply $G_2(c_2, s_2)$ to H, where

$$
(3.15) \qquad c_2 = h_{22}/\sqrt{h_{22}^2 + h_{32}^2} \quad \text{and} \quad s_2 = -h_{32}/\sqrt{h_{22}^2 + h_{32}^2}.
$$

and annihilate h_{32}. Note that G_2 does not affect the first column of H. Continuing in this way and setting

$$
Q = G_N \ldots G_1
$$

we see that $QH = R$ is upper triangular.

A straightforward application of these ideas to Algorithm gmres would solve the least squares problem by computing the product of k Givens rotations Q, setting $g = \beta Q e_1$, and noting that

$$
\|\beta e_1 - H_k y^k\|_{R^{k+1}} = \|Q(\beta e_1 - H_k y^k)\|_{R^{k+1}} = \|g - R_k y^k\|_{R^{k+1}},
$$

where R_k is the $k + 1 \times k$ triangular factor of the QR factorization of H_k.

In the context of GMRES iteration, however, we can incrementally perform the QR factorization of H as the GMRES iteration progresses [167]. To see this, note that if $R_k = Q_k H_k$ and, after orthogonalization, we add the new column h_{k+2} to H_k, we can update both Q_k and R_k by first multiplying h_{k+2} by Q_k (that is applying the first k Givens rotations to h_{k+2}), then computing the Givens rotation G_{k+1} that annihilates the $(k+2)$nd element of $Q_k h_{k+2}$, and finally setting $Q_{k+1} = G_{k+1} Q_k$ and forming R_{k+1} by augmenting R_k with $G_{k+1} Q_k h_{k+2}$.

The MATLAB implementation of Algorithm gmres stores Q_k by storing the sequences $\{c_j\}$ and $\{s_j\}$ and then computing the action of Q_k on a vector $x \in R^{k+1}$ by applying $G_j(c_j, s_j)$ in turn to obtain

$$Q_k x = G_k(c_k, s_k) \ldots G_1(c_1, s_1) x.$$

We overwrite the upper triangular part of H_k with the triangular part of the QR factorization of H_k in the MATLAB code. The MATLAB implementation of Algorithm gmres uses (3.10) to test for loss of orthogonality.

ALGORITHM 3.5.1. gmres$(x, b, A, \epsilon, kmax, \rho)$

1. $r = b - Ax$, $v_1 = r/\|r\|_2$, $\rho = \|r\|_2$, $\beta = \rho$,
   $k = 0$; $g = \rho(1, 0, \ldots, 0)^T \in R^{kmax+1}$

2. While $\rho > \epsilon \|b\|_2$ and $k < kmax$ do

   (a) $k = k + 1$

   (b) $v_{k+1} = Av_k$
       for $j = 1, \ldots k$
       i. $h_{jk} = v_{k+1}^T v_j$
       ii. $v_{k+1} = v_{k+1} - h_{jk} v_j$

   (c) $h_{k+1,k} = \|v_{k+1}\|_2$

   (d) Test for loss of orthogonality and reorthogonalize if necessary.

   (e) $v_{k+1} = v_{k+1}/\|v_{k+1}\|_2$

   (f) i. If $k > 1$ apply Q_{k-1} to the kth column of H.
        ii. $\nu = \sqrt{h_{k,k}^2 + h_{k+1,k}^2}$.
        iii. $c_k = h_{k,k}/\nu$, $s_k = -h_{k+1,k}/\nu$
             $h_{k,k} = c_k h_{k,k} - s_k h_{k+1,k}$, $h_{k+1,k} = 0$
        iv. $g = G_k(c_k, s_k)g$.

   (g) $\rho = |(g)_{k+1}|$.

3. Set $r_{i,j} = h_{i,j}$ for $1 \le i, j \le k$.
   Set $(w)_i = (g)_i$ for $1 \le i \le k$.
   Solve the upper triangular system $Ry^k = w$.

4. $x_k = x_0 + V_k y^k$.

We close with an example of an implementation of GMRES(m) . This implementation does not test for success and may, therefore, fail to terminate. You are asked to repair this in exercise 7. Aside from the parameter m, the arguments to Algorithm gmresm are the same as those for Algorithm gmres.

ALGORITHM 3.5.2. $\text{gmresm}(x, b, A, \epsilon, kmax, m, \rho)$

1. $\text{gmres}(x, b, A, \epsilon, m, \rho)$

2. $k = m$

3. While $\rho > \epsilon \|b\|_2$ and $k < kmax$ do

   (a) $\text{gmres}(x, b, A, \epsilon, m, \rho)$

   (b) $k = k + m$

The storage costs of m iterations of gmres or of gmresm are the $m + 2$ vectors b, x, and $\{v_k\}_{k=1}^m$.

3.6.   Other methods for nonsymmetric systems

The methods for nonsymmetric linear systems that receive most attention in this book, GMRES, CGNR, and CGNE, share the properties that they are easy to implement, can be analyzed by a common residual polynomial approach, and only terminate if an acceptable approximate solution has been found. CGNR and CGNE have the disadvantages that a transpose-vector product must be computed for each iteration and that the coefficient matrix $A^T A$ or $A A^T$ has condition number the square of that of the original matrix. In § 3.8 we give an example of how this squaring of the condition number can lead to failure of the iteration. GMRES needs only matrix-vector products and uses A alone, but, as we have seen, a basis for $\mathcal{K}_k$ must be stored to compute x_k. Hence, the storage requirements increase as the iteration progresses. For large and ill-conditioned problems, it may be impossible to store enough basis vectors and the iteration may have to be restarted. Restarting can seriously degrade performance.

An ideal method would, like CG, only need matrix-vector products, be based on some kind of minimization principle or conjugacy, have modest storage requirements that do not depend on the number of iterations needed for convergence, and converge in N iterations for all nonsingular A. However, [74], methods based on short-term recurrences such as CG that also satisfy minimization or conjugacy conditions cannot be constructed for general matrices. The methods we describe in this section fall short of the ideal, but can still be quite useful. We discuss only a small subset of these methods and refer the reader to [12] and [78] for pointers to more of the literature on this subject. All the methods we present in this section require two matrix-vector products for each iteration.

Consistently with our implementation of GMRES, we take the view that preconditioners will be applied externally to the iteration. However, as with CG, these methods can also be implemented in a manner that incorporates the preconditioner and uses the residual for the original system to control termination.

3.6.1. Bi-CG. The earliest such method Bi-CG (Biconjugate gradient) [122], [76], does not enforce a minimization principle; instead, the kth residual must satisfy the *bi-orthogonality condition*

$$(3.16) \qquad r_k^T w = 0 \text{ for all } w \in \overline{\mathcal{K}}_k,$$

where

$$\overline{\mathcal{K}}_k = \text{span}(\hat{r}_0, A^T \hat{r}_0, \ldots, (A^T)^{k-1} \hat{r}_0)$$

is the Krylov space for A^T and the vector $\hat{r}_0$. $\hat{r}_0$ is a user-supplied vector and is often set to r_0. The algorithm gets its name because it produces sequences of residuals $\{r_k\}$, $\{\hat{r}_k\}$ and search directions $\{p_k\}$, $\{\hat{p}_k\}$ such that bi-orthogonality holds, *i. e.* $\hat{r}_k^T r_l = 0$ if $k \neq l$ and the search directions $\{p_k\}$ and $\{\hat{p}_k\}$ satisfy the *bi-conjugacy* property

$$\hat{p}_k^T A p_l = 0 \text{ if } k \neq l.$$

In the symmetric positive definite case (3.16) is equivalent to the minimization principle (2.2) for CG [89].

Using the notation of Chapter 2 and [191] we give an implementation of Bi-CG making the choice $\hat{r}_0 = r_0$. This algorithmic description is explained and motivated in more detail in [122], [76], [78], and [191]. We do not recommend use of Bi-CG and present this algorithm only as a basis for discussion of some of the properties of this class of algorithms.

ALGORITHM 3.6.1. $\text{bicg}(x, b, A, \epsilon, kmax)$
1. $r = b - Ax$, $\hat{r} = r$, $\rho_0 = 1$, $\hat{p} = p = 0$, $k = 0$

2. Do While $\|r\|_2 > \epsilon \|b\|_2$ and $k < kmax$
   (a) $k = k + 1$
   (b) $\rho_k = \hat{r}^T r$, $\beta = \rho_k / \rho_{k-1}$
   (c) $p = r + \beta p$, $\hat{p} = \hat{r} + \beta \hat{p}$
   (d) $v = Ap$
   (e) $\alpha = \rho_k / (\hat{p}^T v)$
   (f) $x = x + \alpha p$
   (g) $r = r - \alpha v$; $\hat{r} = \hat{r} - \alpha A^T \hat{p}$

Note that if $A = A^T$ is spd, Bi-CG produces the same iterations as CG (but computes everything except x twice). If, however, A is not spd, there is no guarantee that ρ_k in step 2b or $\hat{p}^T Ap$, the denominator in step 2e, will not vanish. If either $\rho_{k-1} = 0$ or $\hat{p}^T Ap = 0$ we say that a *breakdown* has taken place. Even if these quantities are nonzero but very small, the algorithm can become unstable or produce inaccurate results. While there are approaches

[80], [78], [148], [79], [81], to avoid some breakdowns in many variations of Bi-CG, there are no methods that both limit storage and completely eliminate the possibility of breakdown [74]. All of the methods we present in this section can break down and should be implemented with that possibility in mind. Once breakdown has occurred, one can restart the Bi-CG iteration or pass the computation to another algorithm, such as GMRES. The possibility of breakdown is small and certainly should not eliminate the algorithms discussed below from consideration if there is not enough storage for GMRES to perform well.

Breakdowns aside, there are other problems with Bi-CG. A transpose-vector product is needed, which at the least will require additional programming and may not be possible at all. The performance of the algorithm can be erratic or even unstable with residuals increasing by several orders of magnitude from one iteration to the next. Moreover, the effort in computing $\hat{r}$ at each iteration is wasted in that $\hat{r}$ makes no contribution to x. However, Bi-CG sometimes performs extremely well and the remaining algorithms in this section represent attempts to capture this good performance and damp the erratic behavior when it occurs.

We can compare the best-case performance of Bi-CG with that of GMRES. Note that there is $\bar{p}_k \in \mathcal{P}_k$ such that both

$$(3.17) \qquad r_k = \bar{p}_k(A)r_0 \text{ and } \hat{r}_k = \bar{p}_k(A^T)\hat{r}_0.$$

Hence, by the minimization property for GMRES

$$\|r_k^{GMRES}\|_2 \leq \|r_k^{Bi-CG}\|_2,$$

reminding us that GMRES, if sufficient storage is available, will always reduce the residual more rapidly than Bi-CG (in terms of iterations, but not necessarily in terms of computational work). One should also keep in mind that a single Bi-CG iteration requires two matrix-vector products and a GMRES iterate only one, but that the cost of the GMRES iteration increases (in terms of floating-point operations) as the iteration progresses.

3.6.2. CGS.   A remedy for one of the problems with Bi-CG is the Conjugate Gradient Squared (CGS) algorithm [180]. The algorithm takes advantage of the fact that (3.17) implies that the scalar product $\hat{r}^T r$ in Bi-CG (step 2b of Algorithm bicg) can be represented without using A^T as

$$r_k^T \hat{r}_k = (\bar{p}_k(A)r_0)^T(\bar{p}_k(A^T)\hat{r}_0) = (\bar{p}_k(A)^2 r_0)^T \hat{r}_0.$$

The other references to A^T can be eliminated in a similar fashion and an iteration satisfying
$$(3.18) \qquad r_k = \bar{p}_k(A)^2 r_0$$

is produced, where $\bar{p}_k$ is the same polynomial produced by Bi-CG and used in (3.17). This explains the name, Conjugate Gradient *Squared*.

The work used in Bi-CG to compute $\hat{r}$ is now used to update x. CGS replaces the transpose-vector product with an additional matrix-vector product and applies the square of the Bi-CG polynomial to r_0 to produce r_k. This may, of course, change the convergence properties for the worse and either improve good convergence or magnify erratic behavior [134], [191]. CGS has the same potential for breakdown as Bi-CG. Since $\bar{p}_k^2 \in \mathcal{P}_{2k}$ we have

$$(3.19) \qquad \|r_{2k}^{GMRES}\|_2 \le \|r_k^{CGS}\|_2.$$

We present an algorithmic description of CGS that we will refer to in our discussion of TFQMR in § 3.6.4. In our description of CGS we will index the vectors and scalars that would be overwritten in an implementation so that we can describe some of the ideas in TFQMR later on. This description is taken from [77].

ALGORITHM 3.6.2. $cgs(x, b, A, \epsilon, kmax)$

1. $x_0 = x;\ p_0 = u_0 = r_0 = b - Ax$

2. $v_0 = Ap_0;\ \hat{r}_0 = r_0$

3. $\rho_0 = \hat{r}_0^T r_0;\ k = 0$

4. Do While $\|r\|_2 > \epsilon \|b\|_2$ and $k < kmax$

   (a) $k = k + 1$

   (b) $\sigma_{k-1} = \hat{r}_0^T v_{k-1}$

   (c) $\alpha_{k-1} = \rho_{k-1}/\sigma_{k-1}$

   (d) $q_k = u_{k-1} - \alpha_{k-1} v_{k-1}$

   (e) $x_k = x_{k-1} + \alpha_{k-1}(u_{k-1} + q_k)$
       $r_k = r_{k-1} - \alpha_{k-1} A(u_{k-1} + q_k)$

   (f) $\rho_k = \hat{r}_0^T r_k;\ \beta_k = \rho_k/\rho_{k-1}$
       $u_k = r_k + \beta_k q_k$
       $p_k = u_k + \beta_k(q_k + \beta_k p_{k-1})$
       $v_k = Ap_k$

Breakdowns take place when either ρ_{k-1} or σ_{k-1} vanish. If the algorithm does not break down, then $\alpha_{k-1} \ne 0$ for all k. One can show [180], [77], that if $\bar{p}_k$ is the residual polynomial from (3.17) and (3.18) then

$$(3.20) \qquad \bar{p}_k(z) = \bar{p}_{k-1}(z) - \alpha_{k-1} z \bar{q}_{k-1}(z)$$

where the auxiliary polynomials $\bar{q}_k \in \mathcal{P}_k$ are given by $\bar{q}_0 = 1$ and for $k \ge 1$ by

$$(3.21) \qquad \bar{q}_k(z) = \bar{p}_k(z) + \beta_k \bar{q}_{k-1}(z).$$

We provide no MATLAB implementation of Bi-CG or CGS. Bi-CGSTAB is the most effective representative of this class of algorithms.

3.6.3. Bi-CGSTAB. The Bi-conjugate gradient stabilized method (Bi-CGSTAB) [191] attempts to smooth the convergence of CGS by replacing (3.18) with

$$(3.22) \qquad r_k = q_k(A)p_k(A)r_0$$

where

$$q_k(z) = \prod_{i=1}^{k}(1 - \omega_i z).$$

The constant ω_i is selected to minimize $r_i = q_i(A)p_i(A)r_0$ as a function of ω_i. The examples in [191] indicate the effectiveness of this approach, which can be thought of [12] as a blend of Bi-CG and GMRES(1). The performance in cases where the spectrum has a significant imaginary part can be improved by constructing q_k to have complex conjugate pairs of roots, [98].

There is no convergence theory for Bi-CGSTAB and no estimate of the residuals that is better than that for CGS, (3.19). In our description of the implementation, which is described and motivated fully in [191], we use $\hat{r}_0 = r_0$ and follow the notation of [191].

ALGORITHM 3.6.3. $\texttt{bicgstab}(x, b, A, \epsilon, kmax)$

1. $r = b - Ax$, $\hat{r}_0 = \hat{r} = r$, $\rho_0 = \alpha = \omega_0 = 1$, $v = p = 0$, $k = 0$, $\rho_1 = \hat{r}_0^T r$

2. Do While $\|r\|_2 > \epsilon\|b\|_2$ and $k < kmax$

   (a) $k = k + 1$

   (b) $\beta = (\rho_k/\rho_{k-1})(\alpha/\omega)$

   (c) $p = r + \beta(p - \omega v)$

   (d) $v = Ap$

   (e) $\alpha = \rho_k/(\hat{r}_0^T v)$

   (f) $s = r - \alpha v$, $t = As$

   (g) $\omega = t^T s/\|t\|_2^2$, $\rho_{k+1} = -\omega \hat{r}_0^T t$

   (h) $x = x + \alpha p + \omega s$

   (i) $r = s - \omega t$

Note that the iteration can break down in steps 2b and 2e. We provide an implementation of Algorithm $\texttt{bicgstab}$ in the collection of MATLAB codes.

The cost in storage and in floating-point operations per iteration remains bounded for the entire iteration. One must store seven vectors $(x, b, r, p, v, \hat{r}_0, t)$, letting s overwrite r when needed. A single iteration requires four scalar products. In a situation where many GMRES iterations are needed and matrix-vector product is fast, Bi-CGSTAB can have a much lower average cost per iterate than GMRES. The reason for this is that the cost of orthogonalization in GMRES can be much more than that of a matrix-vector product if the dimension of the Krylov space is large. We will present such a case in § 3.8.

3.6.4. TFQMR. Now we consider the quasi-minimal residual (QMR) family of algorithms [80], [81], [77], [37], [202]. We will focus on the transpose-free algorithm TFQMR proposed in [77], to illustrate the quasi-minimization idea. All algorithms in this family minimize the norm of an easily computable quantity $q = f - Hz$, a *quasi-residual* over $z \in R^N$. The quasi-residual is related to the true residual by a full-rank linear transformation $r = Lq$ and reduction of the residual can be approximately measured by reduction in q. The specific formulation of q and L depends on the algorithm.

Returning to Algorithm cgs we define sequences

$$
y_m = \begin{cases} u_{k-1} & \text{if } m = 2k - 1 \text{ is odd,} \\ \\ q_k & \text{if } m = 2k \text{ is even,} \end{cases}
$$

and

$$
w_m = \begin{cases} (\bar{p}_k(A))^2 r_0 & \text{if } m = 2k - 1 \text{ is odd,} \\ \\ \bar{p}_k(A)\bar{p}_{k-1}(A)r_0 & \text{if } m = 2k \text{ is even.} \end{cases}
$$

Here the sequences $\{\bar{p}_k\}$ and $\{\bar{q}_k\}$ are given by (3.20) and (3.21). Hence, the CGS residual satisfies

$$
r_k^{CGS} = w_{2k+1}.
$$

We assume that CGS does not break down and, therefore, $\alpha_k \neq 0$ for all $k \geq 0$. If we let $\lfloor r \rfloor$ be the nearest integer less than or equal to a real r we have

(3.23) $$ Ay_m = (w_m - w_{m+1})/\alpha_{\lfloor (m-1)/2 \rfloor}, $$

where the denominator is nonzero by assumption. We express (3.23) in matrix form as

(3.24) $$ AY_m = W_{m+1}B_m, $$

where Y_m is the $N \times m$ matrix with columns $\{y_j\}_{j=1}^{m}$, W_{m+1} the $N \times (m+1)$ matrix with columns $\{w_j\}_{j=1}^{m+1}$, and B_m is the $(m+1) \times m$ matrix given by

$$
B_m = \begin{pmatrix} 1 & 0 & \dots & 0 \\ -1 & 1 & \ddots & \vdots \\ 0 & \vdots & \vdots & 0 \\ \vdots & \ddots & -1 & 1 \\ 0 & \dots & 0 & -1 \end{pmatrix} \operatorname{diag}(\alpha_0, \alpha_0, \dots, \alpha_{\lfloor (m-1)/2 \rfloor})^{-1}.
$$

Our assumptions that no breakdown takes place imply that $\mathcal{K}_m$ is the span of $\{y_j\}_{j=1}^{m}$ and hence

(3.25) $$ x_m = x_0 + Y_m z $$

for some $z \in R^m$. Therefore,

(3.26) $$ r_m = r_0 - AY_m z = W_{m+1}(e_1 - B_m z), $$

where, as in our discussion of GMRES, $e_1 = (1, 0, \ldots, 0)^T \in R^m$. In a sense, the conditioning of the matrix W_{m+1} can be improved by scaling. Let

$$(3.27) \qquad \Omega = \text{diag}(\omega_1, \omega_2, \ldots, \omega_m)$$

and rewrite (3.26) as

$$(3.28) \qquad r_m = r_0 - AY_m z = W_{m+1} \Omega_{m+1}^{-1}(f_{m+1} - H_m z)$$

where $f_{m+1} = \Omega_{m+1} e_1 = \omega_{m+1} e_1$ and $H_m = \Omega_{m+1} B_m$ is bidiagonal and hence upper Hessenberg. The diagonal entries in the matrix Ω_{m+1} are called weights. The weights make it possible to derive straightforward estimates of the residual in terms of easily computed terms in the iteration.

If $W_{m+1} \Omega_{m+1}^{-1}$ were an orthogonal matrix, we could minimize r_m by solving the least squares problem

$$(3.29) \qquad \text{minimize}_{z \in R^m} \| f_{m+1} - H_m z \|_2.$$

The quasi-minimization idea is to solve (3.29) (thereby *quasi-minimizing* the residual) to obtain z_m and then set

$$(3.30) \qquad x_m = x_0 + Y_m z_m.$$

Note that if k corresponds to the iteration counter for CGS, each k produces two approximate solutions (corresponding to $m = 2k - 1$ and $m = 2k$).

The solution to the least squares problem (3.29) can be done by using Givens rotations to update the factorization of the upper Hessenberg matrix H_m [77] and this is reflected in the implementation description below. As one can see from that description, approximate residuals are not computed in Algorithm tfqmr. Instead we have the *quasi-residual norm*

$$\tau_m = \| f_{m+1} - H_m z_m \|_2.$$

Now if we pick the weights $\omega_i = \|w_i\|_2$ then the columns of $W_{m+1} \Omega_{m+1}^{-1}$ have norm one. Hence

$$(3.31) \qquad \|r_m\|_2 = \| W_{m+1} \Omega_{m+1}^{-1} (f_{m+1} - H_m z) \|_2 \leq \tau_m \sqrt{m + 1}.$$

We can, therefore, base our termination criterion on τ_m, and we do this in Algorithm tfqmr.

One can [80], [78], also use the quasi-minimization condition to estimate the TFQMR residual in terms of residual polynomials.

THEOREM 3.6.1. *Let A be an $N \times N$ matrix and let $x_0, b \in R^N$ be given. Assume that the TFQMR iteration does not break down or terminate with the exact solution for k iterations and let $1 \leq m \leq 2k$. Let r_m^{GMRES} be the residual for the mth GMRES iteration. Let ξ_m be the smallest singular value of $W_{m+1} \Omega_{m+1}^{-1}$. Then if $\xi_m > 0$*

$$(3.32) \qquad \tau_m \leq \| r_m^{GMRES} \|_2 / \xi_m.$$

Proof. We may write

$$r_m^{GMRES} = r_0 + Y_m z_m^{GMRES}$$

and obtain

$$\|r_m^{GMRES}\|_2 = \|W_{m+1}\Omega_{m+1}^{-1}(f_{m+1}-H_m z_m^{GMRES})\|_2 \geq \xi_m \|f_{m+1}-H_m z_m^{GMRES}\|_2.$$

Hence, by the quasi-minimization property

$$\|r_m^{GMRES}\|_2 \geq \xi_m \tau_m$$

as desired. □

As a corollary we have a finite termination result and a convergence estimate for diagonalizable matrices similar to Theorem 3.1.3.

COROLLARY 3.6.1. *Let A be an $N \times N$ matrix and let $x_0, b \in R^N$ be given. Then within $(N+1)/2$ iterations the TFQMR iteration will either break down or terminate with the solution.*

We apply Theorem 3.1.3 to (3.32) and use (3.31) to obtain the following result.

THEOREM 3.6.2. *Let $A = V\Lambda V^{-1}$ be a nonsingular diagonalizable matrix. Assume that TFQMR does not break down or terminate with the solution for k iterations. For $1 \leq m \leq 2k$ let ξ_m be the smallest singular value of $W_{m+1}\Omega_{m+1}^{-1}$ and let x_m be given by (3.30). Then, if $\xi_m > 0$,*

$$\frac{\|r_m\|_2}{\|r_0\|_2} \leq \sqrt{m+1}\xi_m^{-1}\kappa(V) \max_{z \in \sigma(A)} |\phi(z)|$$

for all $\phi \in \mathcal{P}_m$.

The implementation below follows [77] with $\hat{r}_0 = r_0$ used throughout.

ALGORITHM 3.6.4. tfqmr$(x, b, A, \epsilon, kmax)$

1. $k = 0$, $w_1 = y_1 = r_0 = b - Ax$, $u_1 = v = Ay_1$, $d = 0$
   $\rho_0 = r_0^T r_0$, $\tau = \|r\|_2$, $\theta = 0$, $\eta = 0$

2. Do While $k < kmax$

   (a) $k = k + 1$

   (b) $\sigma_{k-1} = r_0^T v$, $\alpha = \rho_{k-1}/\sigma_{k-1}$, $y_2 = y_1 - \alpha_{k-1}v$, $u_2 = Ay_2$

   (c) For $j = 1,2$ $(m = 2k - 2 + j)$

   　　i. $w = w - \alpha_{k-1}u_j$

   　　ii. $d = y_j + (\theta^2\eta/\alpha_{k-1})d$

   　　iii. $\theta = \|w\|_2/\tau$, $c = 1/\sqrt{1+\theta^2}$

   　　iv. $\tau = \tau\theta c$, $\eta = c^2\alpha_{k-1}$

   　　v. $x = x + \eta d$

   　　vi. If $\tau\sqrt{m+1} \leq \epsilon\|b\|_2$ terminate successfully

   (d) $\rho_k = r_0^T w$, $\beta = \rho_k/\rho_{k-1}$

(e) $y_1 = w + \beta y_2, u_1 = Ay_1$

(f) $v = u_1 + \beta(u_2 + \beta v)$

Note that y_2 and $u_2 = Ay_2$ need only be computed if the loop in step 2c does not terminate when $j = 1$. We take advantage of this in the MATLAB code `tfqmr` to potentially reduce the matrix-vector product cost by one.

3.7. Examples for GMRES iteration

These examples use the code `gmres` from the collection of MATLAB codes. The inputs, described in the comment lines, are the initial iterate x_0, the right-hand side vector b, a MATLAB function for the matrix-vector product, and iteration parameters to specify the maximum number of iterations and the termination criterion. We do not pass the preconditioner to the GMRES code, rather we pass the *preconditioned problem*. In all the GMRES examples, we limit the number of GMRES iterations to 60. For methods like GMRES(m), Bi-CGSTAB, CGNR, and TFQMR, whose storage requirements do not increase with the number of iterations, we allow for more iterations.

We consider the discretization of the partial differential equation

$$(Lu)(x,y) = -(u_{xx}(x,y) + u_{yy}(x,y)) + a_1(x,y)u_x(x,y)$$

(3.33)

$$+a_2(x,y)u_y(x,y) + a_3(x,y)u(x,y) = f(x,y)$$

on $0 < x, y < 1$ subject to homogeneous Dirichlet boundary conditions

$$u(x,0) = u(x,1) = u(0,y) = u(1,y) = 0, \quad 0 < x,y < 1.$$

For general coefficients $\{a_j\}$ the operator L is not self-adjoint and its discretization is not symmetric. As in § 2.7 we discretize with a five-point centered difference scheme with n^2 points and mesh width $h = 1/(n+1)$. The unknowns are

$$u_{ij} \approx u(x_i, x_j)$$

where $x_i = ih$ for $1 \leq i \leq n$. We compute Lu in a matrix-free manner as we did in § 2.7. We let $n = 31$ to create a system with 961 unknowns. As before, we expect second-order accuracy and terminate the iteration when $\|r_k\|_2 \leq h^2\|b\|_2 \approx 9.8 \times 10^{-4}\|b\|_2$.

As a preconditioner we use the fast Poisson solver `fish2d` described in § 2.7. The motivation for this choice is similar to that for the conjugate gradient computations and has been partially explained theoretically in [33], [34], [121], and [139]. If we let Gu denote the action of the Poisson solver on u, the preconditioned system is $GLu = Gf$.

For the computations reported in this section we took

$$a_1(x,y) = 1, a_2(x,y) = 20y, \text{ and } a_3(x,y) = 1.$$

$u_0 = 0$ was the initial iterate. As in § 2.7 we took the right hand side so that the exact solution was the discretization of

$$10xy(1-x)(1-y)\exp(x^{4.5}).$$

In Figs. 3.1 and 3.2 we plot iteration histories corresponding to preconditioned or unpreconditioned GMRES. For the restarted iteration, we restarted the algorithm after three iterations, GMRES(3) in the terminology of § 3.4. Note that the plots are semilog plots of $\|r\|_2/\|b\|_2$ and that the right-hand sides in the preconditioned and unpreconditioned cases are *not* the same. In both figures the solid line is the history of the preconditioned iteration and the dashed line that of the unpreconditioned. Note the importance of preconditioning. The MATLAB `flops` command indicates that the unpreconditioned iteration required 7.6 million floating point operations and converged in 56 iterations. The preconditioned iteration required 1.3 million floating point operations and terminated after 8 iterations. Restarting after 3 iterations is more costly in terms of the iteration count than not restarting. However, even in the unpreconditioned case, the restarted iteration terminated successfully. In this case the MATLAB `flops` command indicates that the unpreconditioned iteration terminated after 223 iterations and 9.7 million floating point operations and the preconditioned iteration required 13 iterations and 2.6 million floating point operations.

The execution times in our environment indicate that the preconditioned iterations are even faster than the difference in operation counts indicates. The reason for this is that the FFT routine is a MATLAB intrinsic and is not interpreted MATLAB code. In other computing environments preconditioners which vectorize or parallelize particularly well might perform in the same way.

3.8. Examples for CGNR, Bi-CGSTAB, and TFQMR iteration

When a transpose is available, CGNR (or CGNE) is an alternative to GMRES that does not require storage of the iteration history. For elliptic problems like (3.33), the transpose (adjoint) of L is given by

$$L^*u = -u_{xx} - u_{yy} - a_1u_x - a_2u_y + a_3u$$

as one can derive by integration by parts. To apply CGNR to the preconditioned problem, we require the transpose of GL, which is given by

$$(GL)^*u = L^*G^*u = L^*Gu.$$

The cost of the application of the transpose of L or GL is the same as that of the application of L or GL itself. Hence, in the case where the cost of the matrix-vector product dominates the computation, a single CGNR iteration costs roughly the same as two GMRES iterations.

However, CGNR has the effect of squaring the condition number, this effect has serious consequences as Fig. 3.3 shows. The dashed line corresponds to

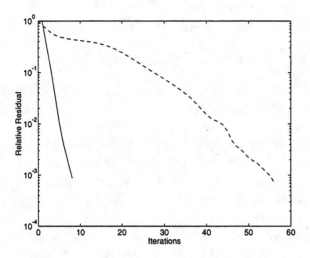

FIG. 3.1. *GMRES for* 2-D *elliptic equation.*

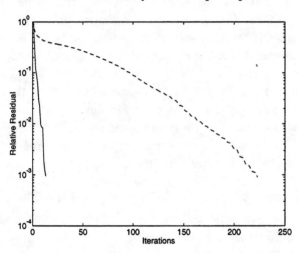

FIG. 3.2. *GMRES*(3) *for* 2-D *elliptic equation.*

CGNR on the unpreconditioned problem, i.e., CG applied to $L^*Lu = L^*f$. The matrix corresponding to the discretization of L^*L has a large condition number. The solid line corresponds to the CGNR applied to the preconditioned problem, i.e., CG applied to $L^*G^2Lu = L^*G^2f$. We limited the number of iterations to $310 = 10m$ and the unpreconditioned formulation of CGNR had made very little progress after 310 iterations. This is a result of squaring the large condition number. The preconditioned formulation did quite well, converging in eight iterations. The MATLAB flops command indicates that the unpreconditioned iteration required 13.7 million floating-point operations and the preconditioned iteration 2.4 million. However, as you are asked to investigate in Exercise 3.9.8, the behavior of even the preconditioned iteration can also be poor if the coefficients of the first derivative terms are too large. CGNE has similar performance; see Exercise 3.9.9.

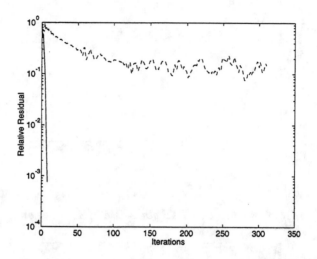

FIG. 3.3.   *CGNR for* 2-D *elliptic equation.*

The examples for Bi-CGSTAB use the code `bicgstab` from the collection of MATLAB codes. This code has the same input/output arguments as `gmres`. We applied Bi-CGSTAB to both the preconditioned and unpreconditioned forms of (3.33) with the same data and termination criterion as for the GMRES example.

We plot the iteration histories in Fig. 3.4. As in the previous examples, the solid line corresponds to the preconditioned iteration and the dashed line to the unpreconditioned iteration. Neither convergence history is monotonic, with the unpreconditioned iteration being very irregular, but not varying by many orders of magnitude as a CGS or Bi-CG iteration might. The preconditioned iteration terminated in six iterations (12 matrix-vector products) and required roughly 1.7 million floating-point operations. The unpreconditioned iteration took 40 iterations and 2.2 million floating-point operations. We see a considerable improvement in cost over CGNR and, since our unpreconditioned matrix-vector product is so inexpensive, a better cost/iterate than GMRES in the unpreconditioned case.

We repeat the experiment with TFQMR as our linear solver. We use the code `tfqmr` that is included in our collection. In Fig. 3.5 we plot the convergence histories in terms of the approximate residual $\sqrt{2k+1}\tau_{2k}$ given by (3.31). We plot only full iterations ($m = 2k$) except, possibly, for the final iterate. The approximate residual is given in terms of the quasi-residual norm τ_m rather than the true residual, and the graphs are monotonic. Our

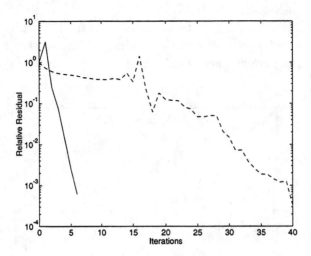

FIG. 3.4. *Bi-CGSTAB for* 2-D *elliptic equation.*

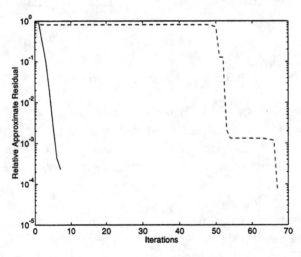

FIG. 3.5. *TFQMR for* 2-D *elliptic equation.*

termination criterion is based on the quasi-residual and (3.31) as described in § 3.6.4, stopping the iteration when $\tau_m \sqrt{m+1}/\|b\|_2 \leq h^{-2}$. This led to actual relative residuals of 7×10^{-5} for the unpreconditioned iteration and 2×10^{-4} for the preconditioned iteration, indicating that it is reasonable to use the quasi-residual estimate (3.31) to make termination decisions.

The unpreconditioned iteration required roughly 4.1 million floating-point operations and 67 iterations for termination. The preconditioned iteration was much more efficient, taking 1.9 million floating-point operations and 7 iterations. The plateaus in the graph of the convergence history for the unpreconditioned iteration are related (see [196]) to spikes in the corresponding graph for a CGS iteration.

3.9.   Exercises on GMRES

3.9.1. Give an example of a matrix that is not diagonalizable.

3.9.2. Prove Theorem 3.1.4 and Theorem 3.1.5.

3.9.3. Prove that the Arnoldi process (unless it terminates prematurely with a solution) produces a basis for $\mathcal{K}_{k+1}$ that satisfies (3.8).

3.9.4. Duplicate Table 3.1 and add two more columns corresponding to classical Gram–Schmidt orthogonalization with reorthogonalization at every step and determine when the test in (3.10) detects loss of orthogonality. How do your conclusions compare with those in [160]?

3.9.5. Let $A = J_\lambda$ be an elementary Jordan block of size N corresponding to an eigenvalue $\lambda \neq 0$. This means that

$$
J_\lambda = \begin{pmatrix}
\lambda & 1 & 0 & & & & \\
0 & \lambda & 1 & 0 & & & \\
 & \ddots & \ddots & \ddots & \ddots & & \\
 & & 0 & \lambda & 1 & 0 \\
 & & & 0 & \lambda & 1 \\
 & & & & 0 & \lambda
\end{pmatrix}.
$$

Give an example of $x_0, b \in R^N$, and $\lambda \in R$ such that the GMRES residuals satisfy $\|r_k\|_2 = \|r_0\|_2$ for all $k < N$.

3.9.6. Consider the centered difference discretization of

$$
-u'' + u' + u = 1, \; u(0) = u(1) = 0.
$$

Solve this problem with GMRES (without preconditioning) and then apply as a preconditioner the map M defined by

$$
-(Mf)'' = f, \; (Mf)(0) = (Mf)(1) = 0.
$$

That is, precondition with a solver for the high-order term in the differential equation using the correct boundary conditions. Try this for meshes with $50, 100, 200, \ldots$ points. How does the performance of the iteration depend on the mesh? Try other preconditioners such as Gauss–Seidel and Jacobi. How do other methods such as CGNR, CGNE, Bi-CGSTAB, and TFQMR perform?

3.9.7. Modify the MATLAB GMRES code to allow for restarts. One way to do this is to call the gmres code with another code which has an outer loop to control the restarting as in Algorithm gmresm. See [167] for another example. How would you test for failure in a restarted algorithm? What kinds of failures can occur? Repeat Exercise 6 with a limit of three GMRES iterations before a restart.

3.9.8. Duplicate the results in § 3.7 and § 3.8. Compare the execution times in your computing environment. Experiment with different values of m. Why might GMRES(m) be faster than GMRES? Vary the dimension and the coefficient a_y. Try $a_y = 5y$. Why is the performance of the methods sensitive to a_y and a_x?

3.9.9. Apply CGNE to the PDE in § 3.7. How does the performance differ from CGNR and GMRES?

3.9.10. Consider preconditioning from the right. For the PDE in § 3.7, apply GMRES, CGNR, Bi-CGSTAB, TFQMR, and/or CGNE, to the equation $LGw = f$ and then recover the solution from $u = Gw$. Compare timings and solution accuracy with those obtained by left preconditioning.

3.9.11. Are the solutions to (3.33) in § 3.7 and § 3.8 equally accurate? Compare the final results with the known solution.

3.9.12. Implement Bi-CG and CGS and try them on some of the examples in this chapter.

Nonlinear Equations

Chapter 4

Basic Concepts and Fixed-Point Iteration

This part of the book is about numerical methods for the solution of systems of nonlinear equations. Throughout this part, we will let $\|\cdot\|$ denote a norm on R^N as well as the induced matrix norm.

We begin by setting the notation. We seek to solve

$$(4.1) \qquad\qquad F(x) = 0.$$

Here $F : R^N \to R^N$. We denote the ith component of F by f_i. If the components of f are differentiable at $x \in R^N$ we define the *Jacobian matrix* $F'(x)$ by

$$F'(x)_{ij} = \frac{\partial f_i}{\partial x_j}(x).$$

The Jacobian matrix is the vector analog of the derivative. We may express the fundamental theorem of calculus as follows.

THEOREM 4.0.1. *Let F be differentiable in an open set $\Omega \subset R^N$ and let $x^* \in \Omega$. Then for all $x \in \Omega$ sufficiently near x^**

$$F(x) - F(x^*) = \int_0^1 F'(x^* + t(x - x^*))(x - x^*)\, dt.$$

4.1. Types of convergence

Iterative methods can be classified by the *rate of convergence*.

DEFINITION 4.1.1. *Let $\{x_n\} \subset R^N$ and $x^* \in R^N$. Then*

- $x_n \to x^*$ q-quadratically *if $x_n \to x^*$ and there is $K > 0$ such that*

$$\|x_{n+1} - x^*\| \le K\|x_n - x^*\|^2.$$

- $x_n \to x^*$ q-superlinearly with q-order $\alpha > 1$ *if $x_n \to x^*$ and there is $K > 0$ such that*

$$\|x_{n+1} - x^*\| \le K\|x_n - x^*\|^\alpha.$$

- $x_n \to x^*$ q-superlinearly *if*

$$\lim_{n\to\infty} \frac{\|x_{n+1} - x^*\|}{\|x_n - x^*\|} = 0.$$

- $x_n \to x^*$ q-linearly *with* q-factor $\sigma \in (0,1)$ *if*

$$\|x_{n+1} - x^*\| \le \sigma \|x_n - x^*\|$$

for n sufficiently large.

DEFINITION 4.1.2. *An iterative method for computing x^* is said to be locally (q-quadratically, q-superlinearly, q-linearly, etc.) convergent if the iterates converge to x^* (q-quadratically, q-superlinearly, q-linearly, etc.) given that the initial data for the iteration is sufficiently good.*

Note that a q-superlinearly convergent sequence is also q-linearly convergent with q-factor σ for any $\sigma > 0$. A q-quadratically convergent sequence is q-superlinearly convergent with q-order 2.

In general, a q-superlinearly convergent method is preferable to a q-linearly convergent one if the cost of a single iterate is the same for both methods. We will see that often a method that is more slowly convergent, in terms of the convergence types defined above, can have a cost/iterate so low that the slow iteration is more efficient.

Sometimes errors are introduced into the iteration that are independent of the errors in the approximate solution. An example of this would be a problem in which the nonlinear function F is the output of another algorithm that provides error control, such as adjustment of the mesh size in an approximation of a differential equation. One might only compute F to low accuracy in the initial phases of the iteration to compute a solution and tighten the accuracy as the iteration progresses. The r-type convergence classification enables us to describe that idea.

DEFINITION 4.1.3. *Let $\{x_n\} \subset R^N$ and $x^* \in R^N$. Then $\{x_n\}$ converges to x^* r-(quadratically, superlinearly, linearly) if there is a sequence $\{\xi_n\} \subset R$ converging q-(quadratically, superlinearly, linearly) to zero such that*

$$\|x_n - x^*\| \le \xi_n.$$

We say that $\{x_n\}$ converges r-superlinearly with r-order $\alpha > 1$ if $\xi_n \to 0$ q-superlinearly with q-order α.

4.2. Fixed-point iteration

Many nonlinear equations are naturally formulated as fixed-point problems

$$(4.2) \qquad x = K(x)$$

where K, the fixed-point map, may be nonlinear. A solution x^* of (4.2) is called a *fixed point* of the map K. Such problems are nonlinear analogs of the linear fixed-point problems considered in Chapter 1. In this section we analyze convergence of *fixed-point iteration*, which is given by

$$(4.3) \qquad x_{n+1} = K(x_n).$$

This iterative method is also called *nonlinear Richardson iteration, Picard iteration,* or *the method of successive substitution.*

Before discussing convergence of fixed-point iteration we make two definitions.

DEFINITION 4.2.1. *Let $\Omega \subset R^N$ and let $G : \Omega \to R^M$. G is Lipschitz continuous on Ω with Lipschitz constant γ if*

$$\|G(x) - G(y)\| \leq \gamma \|x - y\|$$

for all $x, y \in \Omega$.

DEFINITION 4.2.2. *Let $\Omega \subset R^N$. $K : \Omega \to R^N$ is a contraction mapping on Ω if K is Lipschitz continuous on Ω with Lipschitz constant $\gamma < 1$.*

The standard result for fixed-point iteration is the *Contraction Mapping Theorem* [9]. Compare the proof to that of Lemma 1.2.1 and Corollary 1.2.1.

THEOREM 4.2.1. *Let Ω be a closed subset of R^N and let K be a contraction mapping on Ω with Lipschitz constant $\gamma < 1$ such that $K(x) \in \Omega$ for all $x \in \Omega$. Then there is a unique fixed point of K, $x^* \in \Omega$, and the iteration defined by (4.3) converges q-linearly to x^* with q-factor γ for all initial iterates $x_0 \in \Omega$.*

Proof. Let $x_0 \in \Omega$. Note that $\{x_n\} \subset \Omega$ because $x_0 \in \Omega$ and $K(x) \in \Omega$ whenever $x \in \Omega$. The sequence $\{x_n\}$ remains bounded since for all $i \geq 1$

$$\|x_{i+1} - x_i\| = \|K(x_i) - K(x_{i-1})\| \leq \gamma \|x_i - x_{i-1}\| \ldots \leq \gamma^i \|x_1 - x_0\|,$$

and therefore

$$\|x_n - x_0\| \quad = \|\textstyle\sum_{i=0}^{n-1} x_{i+1} - x_i\|$$

$$\leq \textstyle\sum_{i=0}^{n-1} \|x_{i+1} - x_i\| \leq \|x_1 - x_0\| \textstyle\sum_{i=0}^{n-1} \gamma^i$$

$$\leq \|x_1 - x_0\|/(1 - \gamma).$$

Now, for all $n, k \geq 0$,

$$\|x_{n+k} - x_n\| \quad = \|K(x_{n+k-1}) - K(x_{n-1})\|$$

$$\leq \gamma \|x_{n+k-1} - x_{n-1}\|$$

$$\leq \gamma \|K(x_{n+k-2}) - K(x_{n-2})\|$$

$$\leq \gamma^2 \|x_{n+k-2} - x_{n-2}\| \leq \ldots \leq \gamma^n \|x_k - x_0\|$$

$$\leq \gamma^n \|x_1 - x_0\|/(1 - \gamma).$$

Hence

$$\lim_{n,k \to \infty} \|x_{n+k} - x_n\| = 0$$

and therefore the sequence $\{x_n\}$ is a *Cauchy sequence* and has a limit x^* [162].

If K has two fixed points x^* and y^* in Ω, then

$$\|x^* - y^*\| = \|K(x^*) - K(y^*)\| \leq \gamma \|x^* - y^*\|.$$

Since $\gamma < 1$ this can only hold if $\|x^* - y^*\| = 0$, *i. e.* $x^* = y^*$. Hence the fixed point is unique.

Finally we note that

$$\|x_{n+1} - x^*\| = \|K(x_n) - K(x^*)\| \leq \gamma \|x_n - x^*\|,$$

which completes the proof. $\square$

One simple application of this theorem is to numerical integration of ordinary differential equations by implicit methods. For example, consider the initial value problem

$$y' = f(y), y(t_0) = y_0$$

and its solution by the backward Euler method with time step h,

(4.4) $$y^{n+1} - y^n = hf(y^{n+1}).$$

In (4.4) y^k denotes the approximation of $y(t_0 + kh)$. Advancing in time from y^n to y^{n+1} requires solution of the fixed-point problem

$$y = K(y) = y^n + hf(y).$$

For simplicity we assume that f is bounded and Lipschitz continuous, say

$$|f(y)| \leq M \text{ and } |f(x) - f(y)| \leq M|x - y|$$

for all x, y.

If $hM < 1$, we may apply the contraction mapping theorem with $\Omega = R$ since for all x, y

$$\|K(x) - K(y)\| = h|f(x) - f(y)| \leq hM|x - y|.$$

From this we conclude that for h sufficiently small we can integrate forward in time and that the time step h can be chosen independently of y. This time step may well be too small to be practical, and the methods based on Newton's method which we present in the subsequent chapters are used much more often [83]. This type of analysis was used by Picard to prove existence of solutions of initial value problems [152]. Nonlinear variations of the classical stationary iterative methods such as Gauss–Seidel have also been used; see [145] for a more complete discussion of these algorithms.

4.3. The standard assumptions

We will make the *standard assumptions* on F.

ASSUMPTION 4.3.1.

1. *Equation 4.1 has a solution* x^*.

2. $F' : \Omega \to R^{N \times N}$ *is Lipschitz continuous with Lipschitz constant* γ.

3. $F'(x^*)$ *is nonsingular.*

These assumptions can be weakened [108] without sacrificing convergence of the methods we consider here. However the classical result on quadratic convergence of Newton's method requires them and we make them throughout. Exercise 5.7.1 illustrates one way to weaken the standard assumptions.

Throughout this part, we will always denote a root of F by x^*. We let $\mathcal{B}(r)$ denote the ball of radius r about x^*

$$\mathcal{B}(r) = \{x \mid \|e\| < r\},$$

where

$$e = x - x^*.$$

The notation introduced above will be used consistently. If x_n is the nth iterate of a sequence, $e_n = x_n - x^*$ is the error in that iterate.

Lemma 4.3.1. is an important consequence of the standard assumptions.

LEMMA 4.3.1. *Assume that the standard assumptions hold. Then there is $\delta > 0$ so that for all $x \in \mathcal{B}(\delta)$*

(4.5)
$$\|F'(x)\| \leq 2\|F'(x^*)\|,$$

(4.6)
$$\|F'(x)^{-1}\| \leq 2\|F'(x^*)^{-1}\|,$$

and
(4.7)
$$\|F'(x^*)^{-1}\|^{-1}\|e\|/2 \leq \|F(x)\| \leq 2\|F'(x^*)\|\|e\|.$$

Proof. Let δ be small enough so that $\mathcal{B}(\delta) \subset \Omega$. For all $x \in \mathcal{B}(\delta)$ we have

$$\|F'(x)\| \leq \|F'(x^*)\| + \gamma\|e\|.$$

Hence (4.5) holds if $\gamma\delta < \|F'(x^*)\|$.

The next result (4.6) is a direct consequence of the Banach Lemma if

$$\|I - F'(x^*)^{-1}F'(x)\| < 1/2.$$

This will follow from

$$\delta < \frac{\|F'(x^*)^{-1}\|^{-1}}{2\gamma}$$

since then

(4.8)
$$\|I - F'(x^*)^{-1}F'(x)\| = \|F'(x^*)^{-1}(F'(x^*) - F'(x))\|$$
$$\leq \gamma\|F'(x^*)^{-1}\|\|e\| \leq \gamma\delta\|F'(x^*)^{-1}\| < 1/2.$$

To prove the final inequality (4.7), we note that if $x \in \mathcal{B}(\delta)$ then $x^* + te \in \mathcal{B}(\delta)$ for all $0 \leq t \leq 1$. We use (4.5) and Theorem 4.0.1 to obtain

$$\|F(x)\| \leq \int_0^1 \|F'(x^* + te)\|\|e\| \, dt \leq 2\|F'(x^*)\|\|e\|$$

which is the right inequality in (4.7).

To prove the left inequality in (4.7) note that

$$F'(x^*)^{-1}F(x) = F'(x^*)^{-1}\int_0^1 F'(x^* + te)e\,dt$$

$$= e - \int_0^1 (I - F'(x^*)^{-1}F'(x^* + te))e\,dt,$$

and hence, by (4.8)

$$\|F'(x^*)^{-1}F(x)\| \geq \|e\|(1 - \|\int_0^1 I - F'(x^*)^{-1}F'(x^* + te)dt\|) \geq \|e\|/2.$$

Therefore
$$\|e\|/2 \leq \|F'(x^*)^{-1}F(x)\| \leq \|F'(x^*)^{-1}\|\|F(x)\|,$$

which completes the proof. □

Chapter 5

Newton's Method

5.1. Local convergence of Newton's method

We will describe iterative methods for nonlinear equations in terms of the transition from a current iterate x_c to a new iterate x_+. In this language, Newton's method is

(5.1) $$x_+ = x_c - F'(x_c)^{-1}F(x_c).$$

We may also view x_+ as the root of the two-term Taylor expansion or linear model of F about x_c

$$M_c(x) = F(x_c) + F'(x_c)(x - x_c).$$

In the context of single systems this method appeared in the 17th century [140], [156], [13], [137].

The convergence result on Newton's method follows from Lemma 4.3.1.

THEOREM 5.1.1. *Let the standard assumptions hold. Then there are $K > 0$ and $\delta > 0$ such that if $x_c \in \mathcal{B}(\delta)$ the Newton iterate from x_c given by (5.1) satisfies*

(5.2) $$\|e_+\| \leq K\|e_c\|^2.$$

Proof. Let δ be small enough so that the conclusions of Lemma 4.3.1 hold. By Theorem 4.0.1

$$e_+ = e_c - F'(x_c)^{-1}F(x_c) = F'(x_c)^{-1}\int_0^1 (F'(x_c) - F'(x^* + te_c))e_c\, dt.$$

By Lemma 4.3.1 and the Lipschitz continuity of F'

$$\|e_+\| \leq (2\|F'(x^*)^{-1}\|)\gamma\|e_c\|^2/2.$$

This completes the proof of (5.2) with $K = \gamma\|F'(x^*)^{-1}\|$. □

The proof of convergence of the complete Newton iteration will be complete if we reduce δ if needed so that $K\delta < 1$.

THEOREM 5.1.2. *Let the standard assumptions hold. Then there is δ such that if $x_0 \in \mathcal{B}(\delta)$ the Newton iteration*

$$x_{n+1} = x_n - F'(x_n)^{-1}F(x_n)$$

converges q-quadratically to x^*.

Proof. Let δ be small enough so that the conclusions of Theorem 5.1.1 hold. Reduce δ if needed so that $K\delta = \eta < 1$. Then if $n \geq 0$ and $x_n \in \mathcal{B}(\delta)$, Theorem 5.1.1 implies that

$$(5.3) \qquad \|e_{n+1}\| \leq K\|e_n\|^2 \leq \eta\|e_n\| < \|e_n\|$$

and hence $x_{n+1} \in \mathcal{B}(\eta\delta) \subset \mathcal{B}(\delta)$. Therefore if $x_n \in \mathcal{B}(\delta)$ we may continue the iteration. Since $x_0 \in \mathcal{B}(\delta)$ by assumption, the entire sequence $\{x_n\} \subset \mathcal{B}(\delta)$. (5.3) then implies that $x_n \to x^*$ q-quadratically. $\square$

The assumption that the initial iterate be "sufficiently near" the solution $(x_0 \in \mathcal{B}(\delta))$ may seem artificial at first look. There are, however, many situations in which the initial iterate is very near the root. Two examples are implicit integration of ordinary differential equations and differential algebraic equations, [105], [83], [16], where the initial iterate is derived from the solution at the previous time step, and solution of discretizations of partial differential equations, where the initial iterate is an interpolation of a solution from a coarser computational mesh [99], [126]. Moreover, when the initial iterate is far from the root and methods such as those discussed in Chapter 8 are needed, local convergence results like those in this and the following two chapters describe the terminal phase of the iteration.

5.2. Termination of the iteration

We can base our termination decision on Lemma 4.3.1. If $x \in \mathcal{B}(\delta)$, where δ is small enough so that the conclusions of Lemma 4.3.1 hold, then if $F'(x^*)$ is well conditioned, we may terminate the iteration when the *relative nonlinear residual* $\|F(x)\|/\|F(x_0)\|$ is small. We have, as a direct consequence of applying Lemma 4.3.1 twice, a nonlinear form of Lemma 1.1.1

LEMMA 5.2.1. *Assume that the standard assumptions hold. Let $\delta > 0$ be small enough so that the conclusions of Lemma 4.3.1 hold for $x \in \mathcal{B}(\delta)$. Then for all $x \in \mathcal{B}(\delta)$*

$$\frac{\|e\|}{4\|e_0\|\kappa(F'(x^*))} \leq \frac{\|F(x)\|}{\|F(x_0)\|} \leq \frac{4\kappa(F'(x^*))\|e\|}{\|e_0\|},$$

where $\kappa(F'(x^*)) = \|F'(x^*)\|\|F'(x^*)^{-1}\|$ is the condition number of $F'(x^*)$ relative to the norm $\|\cdot\|$.

From Lemma 5.2.1 we conclude that if $F'(x^*)$ is well conditioned, the size of the relative nonlinear residual is a good indicator of the size of the error. However, if there is error in the evaluation of F or the initial iterate is near a root, a termination decision based on the relative residual may be made too late in the iteration or, as we will see in the next section, the iteration may not terminate at all. We raised this issue in the context of linear equations in Chapter 1 when we compared (1.2) and (1.4). In the nonlinear case, there is no "right-hand side" to use as a scaling factor and we must balance the relative and absolute size of the nonlinear residuals in some other way. In all of our MATLAB codes and algorithms for nonlinear equations our termination criterion is to stop the iteration if

$$(5.4) \qquad \qquad \|F(x)\| \leq \tau_r \|F(x_0)\| + \tau_a,$$

where the relative error tolerance τ_r and absolute error tolerance τ_a are input to the algorithm. Combinations of relative and absolute error tolerances are commonly used in numerical methods for ordinary differential equations or differential algebraic equations [16], [21], [23], [151].

Another way to decide whether to terminate is to look at the Newton step

$$s = -F'(x_c)^{-1}F(x_c) = x_+ - x_c,$$

and terminate the iteration when $\|s\|$ is sufficiently small. This criterion is based on Theorem 5.1.1, which implies that

$$\|e_c\| = \|s\| + O(\|e_c\|^2).$$

Hence, near the solution s and e_c are essentially the same size.

For methods other than Newton's method, the relation between the step and the error is less clear. If the iteration is q-linearly convergent, say, then

$$\|e_+\| \leq \sigma \|e_c\|$$

implies that

$$(1 - \sigma)\|e_c\| \leq \|s\| \leq (1 + \sigma)\|e_c\|.$$

Hence the step is a reliable indicator of the error provided σ is not too near 1. The differential equations codes discussed in [16], [21], [23], and [151], make use of this in the special case of the chord method.

However, in order to estimate $\|e_c\|$ this way, one must do most of the work needed to compute x_+, whereas by terminating on small relative residuals or on a condition like (5.4) the decision to stop the iteration can be made before computation and factorization of $F'(x_c)$. If computation and factorization of the Jacobian are inexpensive, then termination on small steps becomes more attractive.

5.3. Implementation of Newton's method

In this section we assume that direct methods will be used to solve the linear equation for the Newton step. Our examples and discussions of operation counts make the implicit assumption that the Jacobian is dense. However, the issues for sparse Jacobians when direct methods are used to solve linear systems are very similar. In Chapter 6 we consider algorithms in which iterative methods are used to solve the equation for the Newton step.

In order to compute the Newton iterate x_+ from a current point x_c one must first evaluate $F(x_c)$ and decide whether to terminate the iteration. If one decides to continue, the Jacobian $F'(x_c)$ must be computed and factored.

Then the step is computed as the solution of $F'(x_c)s = -F(x_c)$ and the iterate is updated $x_+ = x_c + s$. Of these steps, the evaluation and factorization of F' are the most costly. Factorization of F' in the dense case costs $O(N^3)$ floating-point operations. Evaluation of F' by finite differences should be expected to cost N times the cost of an evaluation of F because each column in F' requires an evaluation of F to form the difference approximation. Hence the cost of a Newton step may be roughly estimated as $N + 1$ evaluations of F and $O(N^3)$ floating-point operations. In many cases F' can be computed more efficiently, accurately, and directly than with differences and the analysis above for the cost of a Newton iterate is very pessimistic. See Exercise 5.7.21 for an example. For general nonlinearities or general purpose codes such as the MATLAB code nsol from the collection, there may be little alternative to difference Jacobians. In the future, automatic differentiation (see [94] for a collection of articles on this topic) may provide such an alternative.

For definiteness in the description of Algorithm newton we use an LU factorization of F'. Any other appropriate factorization such as QR or Cholesky could be used as well. The inputs of the algorithm are the initial iterate x, the nonlinear map F, and a vector of termination tolerances $\tau = (\tau_r, \tau_a) \in R^2$.

ALGORITHM 5.3.1. newton(x, F, τ)

1. $r_0 = \|F(x)\|$

2. Do while $\|F(x)\| > \tau_r r_0 + \tau_a$

    (a) Compute $F'(x)$

    (b) Factor $F'(x) = LU$.

    (c) Solve $LUs = -F(x)$

    (d) $x = x + s$

    (e) Evaluate $F(x)$.

One approach to reducing the cost of items 2a and item 2b in the Newton iteration is to move them outside of the main loop. This means that the linear approximation of $F(x) = 0$ that is solved at each iteration has derivative determined by the initial iterate.

$$x_+ = x_c - F'(x_0)^{-1}F(x_c).$$

This method is called the *chord method*. The inputs are the same as for Newton's method.

ALGORITHM 5.3.2. chord(x, F, τ)

1. $r_0 = \|F(x)\|$

2. Compute $F'(x)$

3. Factor $F'(x) = LU$.

4. Do while $\|F(x)\| > \tau_r r_0 + \tau_a$

    (a) Solve $LUs = -F(x)$

    (b) $x = x + s$

    (c) Evaluate $F(x)$.

The only difference in implementation from Newton's method is that the computation and factorization of the Jacobian are done before the iteration is begun. The difference in the iteration itself is that an approximation to $F'(x_c)$ is used. Similar differences arise if $F'(x_c)$ is numerically approximated by differences. We continue in the next section with an analysis of the effects of errors in the function and Jacobian in a general context and then consider the chord method and difference approximation to F' as applications.

5.4. Errors in the function and derivative

Suppose that F and F' are computed inaccurately so that $F + \epsilon$ and $F' + \Delta$ are used instead of F and F' in the iteration. If Δ is sufficiently small, the resulting iteration can return a result that is an $O(\epsilon)$ accurate approximation to x^*. This is different from convergence and was called "local improvement" in [65]. These issues are discussed in [201] as well. If, for example, ϵ is entirely due to floating-point roundoff, there is no reason to expect that $\|F(x_n)\|$ will ever be smaller than ϵ in general. We will refer to this phase of the iteration in which the nonlinear residual is no longer being reduced as *stagnation* of the iteration.

THEOREM 5.4.1. *Let the standard assumptions hold. Then there are $\bar{K} > 0$, $\delta > 0$, and $\delta_1 > 0$ such that if $x_c \in \mathcal{B}(\delta)$ and $\|\Delta(x_c)\| < \delta_1$ then*

$$x_+ = x_c - (F'(x_c) + \Delta(x_c))^{-1}(F(x_c) + \epsilon(x_c))$$

is defined (i.e., $F'(x_c) + \Delta(x_c)$ is nonsingular) and satisfies

$$(5.5) \qquad \|e_+\| \leq \bar{K}(\|e_c\|^2 + \|\Delta(x_c)\|\|e_c\| + \|\epsilon(x_c)\|).$$

Proof. Let δ be small enough so that the conclusions of Lemma 4.3.1 and Theorem 5.1.1 hold. Let

$$x_+^N = x_c - F'(x_c)^{-1}F(x_c)$$

and note that

$$x_+ = x_+^N + (F'(x_c)^{-1} - (F'(x_c) + \Delta(x_c))^{-1})F(x_c) - (F'(x_c) + \Delta(x_c))^{-1}\epsilon(x_c).$$

Lemma 4.3.1 and Theorem 5.1.1 imply

$$\|e_+\| \leq \quad K\|e_c\|^2 + 2\|F'(x_c)^{-1} - (F'(x_c) + \Delta(x_c))^{-1})\|\|F'(x^*)\|\|e_c\|$$
$$(5.6)$$
$$+ \|(F'(x_c) + \Delta(x_c))^{-1}\|\|\epsilon(x_c)\|.$$

If
$$\|\Delta(x_c)\| \leq \|F'(x^*)^{-1}\|^{-1}/4$$
then Lemma 4.3.1 implies that
$$\|\Delta(x_c)\| \leq \|F'(x_c)^{-1}\|^{-1}/2$$
and the Banach Lemma implies that $F'(x_c) + \Delta(x_c)$ is nonsingular and
$$\|(F'(x_c) + \Delta(x_c))^{-1}\| \leq 2\|F'(x_c)^{-1}\| \leq 4\|F'(x^*)^{-1}\|.$$

Hence
$$\|F'(x_c)^{-1} - (F'(x_c) + \Delta(x_c))^{-1}\| \leq 8\|F'(x^*)^{-1}\|^2 \|\Delta(x_c)\|.$$

We use these estimates and (5.6) to obtain
$$\|e_+\| \leq K\|e_c\|^2 + 16\|F'(x^*)^{-1}\|^2\|F'(x^*)\|\|\Delta(x_c)\|\|e_c\| + 4\|F'(x^*)^{-1}\|\|\epsilon(x_c)\|.$$

Setting
$$\bar{K} = K + 16\|F'(x^*)^{-1}\|^2\|F'(x^*)\| + 4\|F'(x^*)^{-1}\|$$
completes the proof. □

The remainder of this section is devoted to applications of Theorem 5.4.1.

5.4.1. The chord method.

Recall that the chord method is given by
$$x_+ = x_c - F'(x_0)^{-1}F(x_c).$$
In the language of Theorem 5.4.1
$$\epsilon(x_c) = 0, \Delta(x_c) = F'(x_0) - F'(x_c).$$

Hence, if $x_c, x_0 \in \mathcal{B}(\delta) \subset \Omega$

(5.7) $$\|\Delta(x_c)\| \leq \gamma\|x_0 - x_c\| \leq \gamma(\|e_0\| + \|e_c\|).$$

We apply Theorem 5.4.1 to obtain the following result.

THEOREM 5.4.2. *Let the standard assumptions hold. Then there are* $K_C > 0$ *and* $\delta > 0$ *such that if* $x_0 \in \mathcal{B}(\delta)$ *the chord iterates converge q-linearly to* x^* *and*

(5.8) $$\|e_{n+1}\| \leq K_C\|e_0\|\|e_n\|.$$

Proof. Let δ be small enough so that $\mathcal{B}(\delta) \subset \Omega$ and the conclusions of Theorem 5.4.1 hold. Assume that $x_n \in \mathcal{B}(\delta)$. Combining (5.7) and (5.5) implies
$$\|e_{n+1}\| \leq \bar{K}(\|e_n\|(1 + \gamma) + \gamma\|e_0\|)\|e_n\| \leq \bar{K}(1 + 2\gamma)\delta\|e_n\|.$$

Hence if δ is small enough so that
$$\bar{K}(1 + 2\gamma)\delta = \eta < 1$$

then the chord iterates converge q-linearly to x^*. Q-linear convergence implies that $\|e_n\| < \|e_0\|$ and hence (5.8) holds with $K_C = \bar{K}(1 + 2\gamma)$. $\square$

Another variation of the chord method is

$$x_+ = x_c - A^{-1}F(x_c),$$

where $A \approx F'(x^*)$. Methods of this type may be viewed as preconditioned nonlinear Richardson iteration. Since

$$\|\Delta(x_c)\| = \|A - F'(x_c)\| \le \|A - F'(x^*)\| + \|F'(x^*) - F'(x_c)\|,$$

if $x_c \in \mathcal{B}(\delta) \subset \Omega$ then

$$\|\Delta(x_c)\| \le \|A - F'(x^*)\| + \gamma\|e_c\| \le \|A - F'(x^*)\| + \gamma\delta.$$

THEOREM 5.4.3. *Let the standard assumptions hold. Then there are $K_A > 0$, $\delta > 0$, and $\delta_1 > 0$, such that if $x_0 \in \mathcal{B}(\delta)$ and $\|A - F'(x^*)\| < \delta_1$ then the iteration*

$$x_{n+1} = x_n - A^{-1}F(x_n)$$

converges q-linearly to x^* and

(5.9) $$\|e_{n+1}\| \le K_A(\|e_0\| + \|A - F'(x^*)\|)\|e_n\|.$$

5.4.2. Approximate inversion of F'.

Another way to implement chord-type methods is to provide an approximate inverse of F'. Here we replace $F'(x)^{-1}$ by $B(x)$, where the action of B on a vector is less expensive to compute than a solution using the LU factorization. Rather than express the iteration in terms of $B^{-1}(x) - F'(x)$ and using Theorem 5.4.3 one can proceed directly from the definition of approximate inverse.

Note that if B is constant (independent of x) then the iteration

$$x_+ = x_c - BF(x_c)$$

can be viewed as a preconditioned nonlinear Richardson iteration.

We have the following result.

THEOREM 5.4.4. *Let the standard assumptions hold. Then there are $K_B > 0$, $\delta > 0$, and $\delta_1 > 0$, such that if $x_0 \in \mathcal{B}(\delta)$ and the matrix-valued function $B(x)$ satisfies*

(5.10) $$\|I - B(x)F'(x^*)\| = \rho(x) < \delta_1$$

for all $x \in \mathcal{B}(\delta)$ then the iteration

$$x_{n+1} = x_n - B(x_n)F(x_n)$$

converges q-linearly to x^* and

(5.11) $$\|e_{n+1}\| \le K_B(\rho(x_n) + \|e_n\|)\|e_n\|.$$

Proof. On one hand, this theorem could be regarded as a corollary of the Banach Lemma and Theorem 5.4.1. We give a direct proof.

First, by (5.10) we have

$$\|B(x)\| = \|B(x)F'(x^*)F'(x^*)^{-1}\| \le \|B(x)F'(x^*)\|\|F'(x^*)^{-1}\|$$

(5.12)
$$\le M_B = (1 + \delta_1)\|F'(x^*)^{-1}\|.$$

Using (5.12) and

$$e_+ = e_c - B(x_c)F(x_c) = \int_0^1 (I - B(x_c)F'(x^* + te_c))e_c \, dt$$

$$= (I - B(x_c)F'(x^*))e_c + B(x_c)\int_0^1 (F'(x^*) - F'(x^* + te_c))e_c \, dt$$

we have

$$\|e_+\| \le \rho(x_c)\|e_c\| + M_B\gamma\|e_c\|^2/2.$$

This completes the proof with $K_B = 1 + M_B\gamma/2$. □

5.4.3. The Shamanskii method.

Alternation of a Newton step with a sequence of chord steps leads to a class of *high-order methods*, that is, methods that converge q-superlinearly with q-order larger than 2. We follow [18] and name this method for Shamanskii [174], who considered the finite difference Jacobian formulation of the method from the point of view of efficiency. The method itself was analyzed in [190]. Other high-order methods, especially for problems in one dimension, are described in [190] and [146].

We can describe the transition from x_c to x_+ by

$$y_1 = x_c - F'(x_c)^{-1}F(x_c),$$

(5.13)
$$y_{j+1} = y_j - F'(x_c)^{-1}F(y_j) \text{ for } 1 \le j \le m - 1,$$

$$x_+ = y_m.$$

Note that $m = 1$ is Newton's method and $m = \infty$ is the chord method with $\{y_j\}$ playing the role of the chord iterates.

Algorithmically a second loop surrounds the factor/solve step. The inputs for the Shamanskii method are the same as for Newton's method or the chord method except for the addition of the parameter m. Note that we can overwrite x_c with the sequence of y_j's. Note that we apply the termination test after computation of each y_j.

ALGORITHM 5.4.1. sham(x, F, τ, m)

1. $r_0 = \|F(x)\|$

2. Do while $\|F(x)\| > \tau_r r_0 + \tau_a$

   (a) Compute $F'(x)$

   (b) Factor $F'(x) = LU$.

   (c) for $j = 1, \ldots m$

      i. Solve $LUs = -F(x)$

      ii. $x = x + s$

      iii. Evaluate $F(x)$.

      iv. If $\|F(x)\| \leq \tau_r r_0 + \tau_a$ exit.

The convergence result is a simple application of Theorem 5.4.2.

THEOREM 5.4.5. *Let the standard assumptions hold and let $m \geq 1$ be given. Then there are $K_S > 0$ and $\delta > 0$ such that if $x_0 \in \mathcal{B}(\delta)$ the Shamanskii iterates converge q-superlinearly to x^* with q-order $m + 1$ and*

$$\text{(5.14)} \qquad \|e_{n+1}\| \leq K_S \|e_n\|^{m+1}.$$

Proof. Let δ be small enough so that $\mathcal{B}(\delta) \subset \Omega$ and that the conclusions of Theorem 5.4.2 hold. Then if $x_n \in \mathcal{B}(\delta)$ all the intermediate iterates $\{y_j\}$ are in $\mathcal{B}(\delta)$ by Theorem 5.4.2. In fact, if we set $y_1 = x_n$, (5.8) implies that for $1 \leq j \leq m$

$$\|y_j - x^*\| \leq K_C \|x_n - x^*\| \|y_{j-1} - x^*\| = K_C \|e_n\| \|y_{j-1} - x^*\| \leq \ldots \leq K_C^j \|e_n\|^{j+1}.$$

Hence $x_{n+1} \in \mathcal{B}(\delta)$. Setting $j = m$ in the above inequality completes the proof. $\square$

The advantage of the Shamanskii method over Newton's method is that high q-orders can be obtained with far fewer Jacobian evaluations or factorizations. Optimal choices of m as a function of N can be derived under assumptions consistent with dense matrix algebra [18] by balancing the $O(N^3)$ cost of a matrix factorization with the cost of function and Jacobian evaluation. The analysis in [18] and the expectation that, if the initial iterate is sufficiently accurate, only a small number of iterations will be taken, indicate that the chord method is usually the best option for very large problems. Algorithm nsol is based on this idea and using an idea from [114] computes and factors the Jacobian only until an estimate of the q-factor for the linear convergence rate is sufficiently low.

5.4.4. Difference approximation to F'.

Assume that we compute $F(x) + \epsilon(x)$ instead of $F(x)$ and attempt to approximate the action of F' on a vector by a forward difference. We would do this, for example, when building an approximate Jacobian. Here the jth column of $F'(x)$ is $F'(x)e_j$ where e_j is the unit vector with jth component 1 and other components 0.

A forward difference approximation to $F'(x)w$ would be

$$\frac{F(x + hw) + \epsilon(x + hw) - F(x) - \epsilon(x)}{h}.$$

Assume that $\|\epsilon(x)\| \leq \bar{\epsilon}$. Then

$$F'(x)w - \frac{F(x+hw) + \epsilon(x+hw) - F(x) - \epsilon(x)}{h} = O(h + \bar{\epsilon}/h).$$

The quantity inside the O-term is minimized when $h = \sqrt{\bar{\epsilon}}$. This means, for instance, that very small difference steps h can lead to inaccurate results. In the special case where $\epsilon(x)$ is a result of floating-point roundoff in full precision ($\bar{\epsilon} \approx 10^{-15}$), the analysis here indicates that $h \approx 10^{-7}$ is a reasonable choice. The choice $h \approx 10^{-15}$ in this case can lead to disaster (try it!). Here we have made the implicit assumption that x and w are of roughly the same size. If this is not the case, h should be scaled to reflect that. The choice

$$h = \bar{\epsilon}^{1/2}\|x\|/\|w\|$$

reflects all the discussion above.

A more important consequence of this analysis is that the choice of step size in a forward difference approximation to the action of the Jacobian on a vector must take into account the error in the evaluation of F. This can become very important if part of F is computed from measured data.

If $F(x)$ has already been computed, the cost of a forward difference approximation of $F'(x)w$ is an additional evaluation of F (at the point $x+hw$). Hence the evaluation of a full finite difference Jacobian would cost N function evaluations, one for each column of the Jacobian. Exploitation of special structure could reduce this cost.

The forward difference approximation to the directional derivative is not a linear operator in w. The reason for this is that the derivative has been approximated, and so linearity has been lost. Because of this we must carefully specify what we mean by a difference approximation to the Jacobian. Definition 5.4.1 below specifies the usual choice.

DEFINITION 5.4.1. *Let F be defined in a neighborhood of $x \in R^N$. $(\nabla_h F)(x)$ is the $N \times N$ matrix whose jth column is given by*

$$(\nabla_h F)(x)_j = \begin{cases} \dfrac{F(x + h\|x\|e_j) - F(x)}{h\|x\|} & x \neq 0 \\[2ex] \dfrac{F(he_j) - F(x)}{h} & x = 0 \end{cases}$$

We make a similar definition of the difference approximation of the directional derivative.

DEFINITION 5.4.2. *Let F be defined in a neighborhood of $x \in R^N$ and let*

$w \in R^N$. *We have*

$$(5.15) D_h F(x:w) = \begin{cases} 0, & w = 0, \\ \\ \|w\| \dfrac{F(x + h\|x\|w/\|w\|) - F(x)}{h\|x\|}, & w, x \neq 0, \\ \\ \|w\| \dfrac{F(hw/\|w\|) - F(x)}{h}, & x = 0, w \neq 0. \end{cases}$$

The standard assumptions and the Banach Lemma imply the following result.

LEMMA 5.4.1. *Let the standard assumptions hold. Then there are $\delta, \bar{\epsilon}, \bar{h} > 0$ such that if $x \in \mathcal{B}(\delta)$, $h \leq \bar{h}$, and $\|\epsilon(x)\| \leq \bar{\epsilon}$ for all $x \in \mathcal{B}(\delta)$, then $\nabla_h(F + \epsilon)(x)$ is nonsingular for all $x \in \mathcal{B}(\delta)$ and there is $M_F > 0$ such that*

$$\|F'(x) - \nabla_h(F + \epsilon)(x)\| \leq M_F(h + \bar{\epsilon}/h).$$

If we assume that the forward difference step has been selected so that $h = O(\sqrt{\bar{\epsilon}})$, then $\Delta(x) = O(\sqrt{\bar{\epsilon}})$ by Lemma 5.4.1. Theorem 5.4.1 implies the following result.

THEOREM 5.4.6. *Let the standard assumptions hold. Then there are positive δ, $\bar{\epsilon}$, and K_D such that if $x_c \in \mathcal{B}(\delta)$, $\|\epsilon(x)\| \leq \bar{\epsilon}$ for all $x \in \mathcal{B}(\delta)$, and there is $M_- > 0$ such that*

$$h_c > M_- \sqrt{\bar{\epsilon}}$$

then

$$x_+ = x_c - \nabla_{h_c}(F + \epsilon)(x_c)^{-1}(F(x_c) + \epsilon(x_c))$$

satisfies

$$\|e_+\| \leq K_D(\bar{\epsilon} + (\|e_c\| + h_c)\|e_c\|).$$

Note that Theorem 5.4.6 does not imply that an iteration will converge to x^*. Even if ϵ is entirely due to floating-point roundoff, there is no guarantee that $\epsilon(x_n) \to 0$ as $x_n \to x^*$. Hence, one should expect the sequence $\{\|e_n\|\}$ to *stagnate* and cease to decrease once $\|e_n\| \approx \bar{\epsilon}$. Therefore in the early phases of the iteration, while $\|e_n\| >> \sqrt{\bar{\epsilon}}$, the progress of the iteration will be indistinguishable from Newton's method as implemented with an exact Jacobian. When $\|e_n\| \leq \sqrt{\bar{\epsilon}}$, Theorem 5.4.6 says that $\|e_{n+1}\| = O(\bar{\epsilon})$. Hence one will see quadratic convergence until the error in F admits no further reduction.

A difference approximation to a full Jacobian does not use any special structure of the Jacobian and will probably be less efficient than a hand coded analytic Jacobian. If information on the sparsity pattern of the Jacobian is available it is sometimes possible [47], [42], to evaluate $\nabla_h F$ with far fewer than N evaluations of F. If the sparsity pattern is such that F' can be reduced to a block triangular form, the entire solution process can take advantage of this structure [59].

For problems in which the Lipschitz constant of F' is large, the error in the difference approximation will be large, too. Moreover, $D_h F(x : w)$ is not, in general, a linear function of w and it is usually the case that

$$(\nabla_h F(x))w \neq D_h F(x : w).$$

If $F'(x^*)$ is ill conditioned or the Lipschitz constant of F' is large, the difference between them may be significant and a difference approximation to a directional derivative, which uses the size of w, is likely to be more accurate than $\nabla_h F(x)w$.

While we used full matrix approximations to numerical Jacobians in all the examples reported here, they should be used with caution.

5.4.5. The secant method.

The results in this section are for single equations $f(x) = 0$, where f is a real-valued function of one real variable. We assume for the sake of simplicity that there are no errors in the evaluation of f, i.e., $\epsilon(x) = 0$. The standard assumptions, then, imply that $f'(x^*) \neq 0$. There is no reason to use a difference approximation to f' for equations in a single variable because one can use previously computed data to approximate $f'(x_n)$ by

$$a_n = \frac{f(x_n) - f(x_{n-1})}{x_n - x_{n-1}}.$$

The resulting method is called the *secant method* and is due to Newton [140], [137]. We will prove that the secant method is locally q-superlinearly convergent if the standard assumptions hold.

In order to start, one needs *two* approximations to x^*, x_0 and x_{-1}. The local convergence theory can be easily described by Theorem 5.4.1. Let x_c be the current iterate and x_- the previous iterate. We have $\epsilon(x_c) = 0$ and

(5.16)
$$\Delta(x_c) = \frac{f(x_c) - f(x_-)}{x_c - x_-} - f'(x_c).$$

We have the following theorem.

THEOREM 5.4.7. *Let the standard assumptions hold with $N = 1$. Then there is $\delta > 0$ such that if $x_0, x_{-1} \in \mathcal{B}(\delta)$ and $x_0 \neq x_{-1}$ then the secant iterates converge q-superlinearly to x^*.*

Proof. Let δ be small enough so that $\mathcal{B}(\delta) \subset \Omega$ and the conclusions of Theorem 5.4.1 hold. Assume that $x_{-1}, x_0, \ldots, x_n \in \mathcal{B}(\delta)$. If $x_n = x^*$ for some finite n, we are done. Otherwise $x_n \neq x_{n-1}$. If we set $s = x_n - x_{n-1}$ we have by (5.16) and the fundamental theorem of calculus

$$\Delta(x_n) = \int_0^1 (f'(x_{n-1} + ts) - f'(x_n))\, dt$$

and hence

$$|\Delta(x_n)| \leq \gamma|x_{n-1} - x_n|/2 \leq \frac{\gamma(|e_n| + |e_{n-1}|)}{2}.$$

Applying Theorem 5.4.1 implies

(5.17) $$|e_{n+1}| \leq \bar{K}((1 + \gamma/2)|e_n|^2 + \gamma|e_n||e_{n-1}|/2).$$

If we reduce δ so that

$$\bar{K}(1 + \gamma)\delta = \eta < 1$$

then $x_n \to x^*$ q-linearly. Therefore

$$\frac{|e_{n+1}|}{|e_n|} \leq \bar{K}((1 + \gamma/2)|e_n| + \gamma|e_{n-1}|/2) \to 0$$

as $n \to \infty$. This completes the proof. □

The secant method and the classical bisection (see Exercise 5.7.8) method are the basis for the popular method of Brent [17] for the solution of single equations.

5.5. The Kantorovich Theorem

In this section we present the result of Kantorovich [107], [106], which states that if the standard assumptions "almost" hold at a point x_0, then there is a root and the Newton iteration converges r-quadratically to that root. This result is of use, for example, in proving that discretizations of nonlinear differential and integral equations have solutions that are near to that of the continuous problem.

We require the following assumption.

ASSUMPTION 5.5.1. *There are β, η, $\bar{r}$, and γ with $\beta\eta\gamma \leq 1/2$ and $x_0 \in R^N$ such that*

1. F is differentiable at x_0,

$$\|F'(x_0)^{-1}\| \leq \beta, \ \ and \ \ \|F'(x_0)^{-1}F(x_0)\| \leq \eta.$$

2. F' is Lipschitz continuous with Lipschitz constant γ in a ball of radius $\bar{r} \geq r_-$ about x_0 where

$$r_- = \frac{1 - \sqrt{1 - 2\beta\eta\gamma}}{\beta\gamma}.$$

We will let $\mathcal{B}_0$ be the closed ball of radius r_- about x_0

$$\mathcal{B}_0 = \{x \mid \|x - x_0\| \leq r_-\}.$$

We do not prove the result in its full generality and refer to [106], [145], [57], and [58] for a complete proof and for discussions of related results. However, we give a simplified version of a Kantorovich-like result for the chord method to show how existence of a solution can be derived from estimates on the function and Jacobian. Unlike the chord method results in [106], [145], and [57], Theorem 5.5.1 assumes a bit more than Assumption 5.5.1 but has a simple proof that uses only the contraction mapping theorem and the fundamental theorem of calculus and obtains q-linear convergence instead of r-linear convergence.

THEOREM 5.5.1. *Let Assumption* 5.5.1 *hold and let*

$$(5.18) \qquad\qquad \beta\gamma\eta < 1/2.$$

Then there is a unique root x^* of F in $\mathcal{B}_0$, the chord iteration with x_0 as the initial iterate converges to x^* q-linearly with q-factor $\beta\gamma r_-$, and $x_n \in \mathcal{B}_0$ for all n. Moreover x^* is the unique root of F in the ball of radius

$$\min\left(\bar{r}, \frac{1 + \sqrt{1 - 2\beta\eta\gamma}}{\beta\gamma}\right)$$

about x_0.

Proof. We will show that the map

$$\phi(x) = x - F'(x_0)^{-1}F(x)$$

is a contraction on the set

$$\mathcal{S} = \{x \mid \|x - x_0\| \leq r_-\}.$$

By the fundamental theorem of calculus and Assumption 5.5.1 we have for all $x \in \mathcal{S}$

$$F'(x_0)^{-1}F(x) = F'(x_0)^{-1}F(x_0)$$

$$+ F'(x_0)^{-1} \int_0^1 F'(x_0 + t(x - x_0))(x - x_0)\, dt$$

$$(5.19) \qquad = F'(x_0)^{-1}F(x_0) + x - x_0$$

$$+ F'(x_0)^{-1} \int_0^1 (F'(x_0 + t(x - x_0)) - F'(x_0))(x - x_0)\, dt$$

and

$$\phi(x) - x_0 = -F'(x_0)^{-1}F(x_0)$$

$$- F'(x_0)^{-1} \int_0^1 (F'(x_0 + t(x - x_0)) - F'(x_0))(x - x_0)\, dt.$$

Hence,

$$\|\phi(x) - x_0\| \leq \eta + \beta\gamma r_-^2/2 = r_-.$$

and ϕ maps S into itself.

To show that ϕ is a contraction on S and to prove the q-linear convergence we will show that

$$\|\phi(x) - \phi(y)\| \leq \beta\gamma r_-\|x - y\|$$

for all $x, y \in S$. If we do this, the contraction mapping theorem (Theorem 4.2.1) will imply that there is a unique fixed point x^* of ϕ in S and that $x_n \to x^*$ q-linearly with q-factor $\beta\gamma r_-$. Clearly x^* is a root of F because it is a fixed point of ϕ. We know that $\beta\gamma r_- < 1$ by our assumption that $\beta\eta\gamma < 1/2$.

Note that for all $x \in \mathcal{B}_0$

$$\phi'(x) = I - F'(x_0)^{-1}F'(x) = F'(x_0)^{-1}(F'(x_0) - F'(x))$$

and hence

$$\|\phi'(x)\| \leq \beta\gamma\|x - x_0\| \leq \beta\gamma r_- < 1.$$

Therefore, for all $x, y \in \mathcal{B}_0$,

$$\|\phi(x) - \phi(y)\| = \left\|\int_0^1 \phi'(y + t(x - y))(x - y)\,dt\right\| \leq \beta\gamma r_-\|x - y\|.$$

This proves the convergence result and the uniqueness of x^* in $\mathcal{B}_0$.

To prove the remainder of the uniqueness assertion let

$$r_+ = \frac{1 + \sqrt{1 - 2\beta\eta\gamma}}{\beta\gamma}$$

and note that $r_+ > r_-$ because $\beta\eta\gamma < 1/2$. If $\bar{r} = r_-$, then we are done by the uniqueness of x^* in $\mathcal{B}_0$. Hence we may assume that $\bar{r} > r_-$. We must show that if x is such that

$$r_- < \|x - x_0\| < \min(\bar{r}, r_+)$$

then $F(x) \neq 0$. Letting $r = \|x - x_0\|$ we have by (5.19)

$$\|F'(x_0)^{-1}F(x)\| \geq r - \eta - \beta\gamma r^2/2 > 0$$

because $r_- < r < r_+$. This completes the proof. $\square$

The Kantorovich theorem is more precise than Theorem 5.5.1 and has a very different proof. We state the theorem in the standard way [145], [106], [63], using the r-quadratic estimates from [106].

THEOREM 5.5.2. *Let Assumption 5.5.1 hold. Then there is a unique root x^* of F in $\mathcal{B}_0$, the Newton iteration with x_0 as the initial iterate converges to x^*, and $x_n \in \mathcal{B}_0$ for all n. x^* is the unique root of F in the ball of radius*

$$\min\left(\bar{r}, \frac{1 + \sqrt{1 - 2\beta\eta\gamma}}{\beta\gamma}\right)$$

about x_0 and the errors satisfy

(5.20)
$$\|e_n\| \leq \frac{(2\beta\eta\gamma)^{2^n}}{2^n\beta\gamma}.$$

If $\beta\eta\gamma < 1/2$ then (5.20) implies that the convergence is r-quadratic (see Exercise 5.7.16).

5.6. Examples for Newton's method

In the collection of MATLAB codes we provide an implementation nsol of Newton's method, the chord method, the Shamanskii method, or a combination of all of them based on the reduction in the nonlinear residual. This latter option decides if the Jacobian should be recomputed based on the ratio of successive residuals. If

$$\|F(x_+)\|/\|F(x_c)\|$$

is below a given threshold, the Jacobian is not recomputed. If the ratio is too large, however, we compute and factor $F'(x_+)$ for use in the subsequent chord steps. In addition to the inputs of Algorithm sham, a threshold for the ratio of successive residuals is also input. The Jacobian is recomputed and factored if either the ratio of successive residuals exceeds the threshold $0 < \rho < 1$ or the number of iterations without an update exceeds m. The output of the MATLAB implementation of nsol is the solution, the history of the iteration stored as the vector of l^∞ norms of the nonlinear residuals, and an error flag. nsol uses diffjac to approximate the Jacobian by $\nabla_h F$ with $h = 10^{-7}$.

While our examples have dense Jacobians, the ideas in the implementation of nsol also apply to situations in which the Jacobian is sparse and direct methods for solving the equation for the Newton step are necessary. One would still want to minimize the number of Jacobian computations (by the methods of [47] or [42], say) and sparse matrix factorizations. In Exercise 5.7.25, which is not easy, the reader is asked to create a sparse form of nsol.

ALGORITHM 5.6.1. $\mathrm{nsol}(x, F, \tau, m, \rho)$

1. $r_c = r_0 = \|F(x)\|$

2. Do while $\|F(x)\| > \tau_r r_0 + \tau_a$

    (a) Compute $\nabla_h F(x) \approx F'(x)$

    (b) Factor $\nabla_h F(x) = LU$.

    (c) $i_s = 0$, $\sigma = 0$

    (d) Do While $i_s < m$ and $\sigma \leq \rho$

        i. $i_s = i_s + 1$

        ii. Solve $LUs = -F(x)$

        iii. $x = x + s$

        iv. Evaluate $F(x)$

        v. $r_+ = \|F(x)\|, \sigma = r_+/r_c, r_c = r_+$

        vi. If $\|F(x)\| \leq \tau_r r_0 + \tau_a$ exit.

In the MATLAB code the iteration is terminated with an error condition if $\sigma \geq 1$ at any point. This indicates that the local convergence analysis in this section is not applicable. In Chapter 8 we discuss how to recover from this. nsol is based on the assumption that a difference Jacobian will be used. In Exercise 5.7.21 you are asked to modify the code to accept an analytic Jacobian. Many times the Jacobian can be computed efficiently by reusing results that are already available from the function evaluation. The parameters m and ρ are assigned the default values of 1000 and .5. With these default parameters the decision to update the Jacobian is made based entirely on the reduction in the norm of the nonlinear residual. nsol allows the user to specify a maximum number of iterations; the default is 40.

The Chandrasekhar H-equation.   The Chandrasekhar H-equation, [41], [30],

$$(5.21) \qquad F(H)(\mu) = H(\mu) - \left(1 - \frac{c}{2}\int_0^1 \frac{\mu H(\nu)\,d\nu}{\mu + \nu}\right)^{-1} = 0,$$

is used to solve exit distribution problems in radiative transfer.

We will discretize the equation with the composite midpoint rule. Here we approximate integrals on $[0, 1]$ by

$$\int_0^1 f(\mu)\,d\mu \approx \frac{1}{N}\sum_{j=1}^{N} f(\mu_j),$$

where $\mu_i = (i - 1/2)/N$ for $1 \le i \le N$. The resulting discrete problem is

$$(5.22) \qquad F(x)_i = x_i - \left(1 - \frac{c}{2N}\sum_{j=1}^{N} \frac{\mu_i x_j}{\mu_i + \mu_j}\right)^{-1}.$$

It is known [132] that both (5.21) and the discrete analog (5.22) have two solutions for $c \in (0, 1)$. Only one of these solutions has physical meaning, however. Newton's method, with the 0 function or the constant function with value 1 as initial iterate, will find the physically meaningful solution [110]. The standard assumptions hold near either solution [110] for $c \in [0, 1)$. The discrete problem has its own physical meaning [41] and we will attempt to solve it to high accuracy. Because H has a singularity at $\mu = 0$, the solution of the discrete problem is not even a first-order accurate approximation to that of the continuous problem.

In the computations reported here we set $N = 100$ and $c = .9$. We used the function identically one as initial iterate. We computed all Jacobians with the MATLAB code diffjac. We set the termination parameters $\tau_r = \tau_a = 10^{-6}$. We begin with a comparison of Newton's method and the chord method. We present some *iteration statistics* in both tabular and graphical form. In Table 5.1 we tabulate the iteration counter n, $\|F(x_n)\|_\infty/\|F(x_0)\|_\infty$, and the ratio

$$R_n = \|F(x_n)\|_\infty/\|F(x_{n-1})\|_\infty$$

TABLE 5.1

Comparison of Newton and chord iteration.

n	$\|F(x_n)\|_\infty/\|F(x_0)\|_\infty$	R_n	$\|F(x_n)\|_\infty/\|F(x_0)\|_\infty$	R_n
0	1.000e+00		1.000e+00	
1	1.480e-01	1.480e-01	1.480e-01	1.480e-01
2	2.698e-03	1.823e-02	3.074e-02	2.077e-01
3	7.729e-07	2.865e-04	6.511e-03	2.118e-01
4			1.388e-03	2.132e-01
5			2.965e-04	2.136e-01
6			6.334e-05	2.136e-01
7			1.353e-05	2.136e-01
8			2.891e-06	2.136e-01

for $n \geq 1$ for both Newton's method and the chord method. In Fig. 5.1 the solid curve is a plot of $\|F(x_n)\|_\infty/\|F(x_0)\|_\infty$ for Newton's method and the dashed curve a plot of $\|F(x_n)\|_\infty/\|F(x_0)\|_\infty$ for the chord method. The concave iteration track is the signature of superlinear convergence, the linear track indicates linear convergence. See Exercise 5.7.15 for a more careful examination of this concavity. The linear convergence of the chord method can also be seen in the convergence of the ratios R_n to a constant value.

For this example the MATLAB flops command returns an fairly accurate indicator of the cost. The Newton iteration required over 8.8 million floating-point operations while the chord iteration required under 3.3 million. In Exercise 5.7.18 you are asked to obtain timings in your environment and will see how the floating-point operation counts are related to the actual timings.

The good performance of the chord method is no accident. If we think of the chord method as a preconditioned nonlinear Richardson iteration with the initial Jacobian playing the role of preconditioner, then the chord method should be much more efficient than Newton's method if the equations for the steps can be solved cheaply. In the case considered here, the cost of a single evaluation and factorization of the Jacobian is far more than the cost of the entire remainder of the chord iteration.

The ideas in Chapters 6 and 7 show how the low iteration cost of the chord method can be combined with the small number of iterations of a superlinearly convergent nonlinear iterative method. In the context of this particular example, the cost of the Jacobian computation can be reduced by analytic evaluation and reuse of data that have already been computed for the evaluation of F. The reader is encouraged to do this in Exercise 5.7.21.

In the remaining examples, we plot, but do not tabulate, the iteration statistics. In Fig. 5.2 we compare the Shamanskii method with $m = 2$ with Newton's method. The Shamanskii computation required under 6 million floating-point operations, as compared with over 8.8 for Newton's method. However, for this problem, the simple chord method is the most efficient.

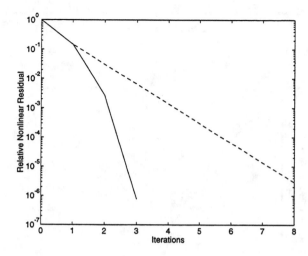

FIG. 5.1. *Newton and chord methods for H-equation, c = .9.*

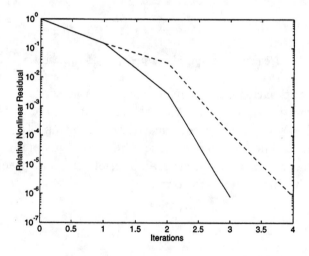

FIG. 5.2. *Newton and Shamanskii method for H-equation, c = .9.*

The hybrid approach in nsol with the default parameters would be the chord method for this problem.

We now reconsider the H-equation with $c = .9999$. For this problem the Jacobian at the solution is nearly singular [110]. Aside from c, all iteration parameters are the same as in the previous example. While Newton's method converges in 7 iterations and requires 21 million floating-point operations, the chord method requires 188 iterations and 10.6 million floating-point operations. The q-factor for the chord method in this computation was over .96, indicating very slow convergence. The hybrid method in nsol with the default parameters computes the Jacobian four times during the course of the iteration, converges in 14 iterations, and requires 12 million floating-point operations. In Fig. 5.3 we compare the Newton iteration with the hybrid.

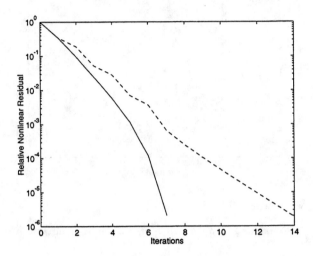

FIG. 5.3. *Newton and hybrid methods for* H-*equation* $c = .9999$.

One should approach plots of convergence histories with some caution. The theory for the chord method asserts that the error norms converge q-linearly to zero. This implies only that the nonlinear residual norms $\{\|F(x_n)\|\}$ converge r-linearly to zero. While the plots indicate q-linear convergence of the nonlinear residuals, the theory does not, in general, support this. In practice, however, plots like those in this section are common.

5.7. Exercises on Newton's method

In some of the exercises here, and in the rest of this part of the book, you will be asked to plot or tabulate *iteration statistics*. When asked to do this, for each iteration tabulate the iteration counter, the norm of F, and the ratio of $\|F(x_n)\|/\|F(x_{n-1})\|$ for $n \geq 1$. A better alternative in the MATLAB environment is to use the `semilogy` command to plot the norms. When one does this, one can visually inspect the plot to determine superlinear convergence (concavity) without explicitly computing the ratios. Use the l^∞ norm.

5.7.1. A function G is said to be *Hölder continuous with exponent α in Ω* if $\|G(x) - G(y)\| \leq K\|x - y\|^\alpha$ for all $x, y \in \Omega$. Show that if the Lipschitz continuity condition on F' in the standard assumptions is replaced by Hölder continuity with exponent $\alpha > 0$ that the Newton iterates converge with q-order $1 + \alpha$.

5.7.2. Can the performance of the Newton iteration be improved by a linear change of variables? That is, for nonsingular $N \times N$ matrices A and B, can the Newton iterates for $F(x) = 0$ and $AF(Bx) = 0$ show any performance difference when started at the same initial iterate? What about the chord method?

5.7.3. Let $\nabla_h F$ be given by Definition 5.4.1. Given the standard assumptions, prove that the iteration given by

$$x_{n+1} = x_n - (\nabla_{h_n} F(x_n))^{-1} F(x_n)$$

is locally q-superlinearly convergent if $h_n \to 0$. When is it locally q-quadratically convergent?

5.7.4. Suppose $F'(x_n)$ is replaced by $\nabla_{h_n} F(x_n)$ in the Shamanskii method. Discuss how h_n must be decreased as the iteration progresses to preserve the q-factor of $m + 1$ [174].

5.7.5. In what way is the convergence analysis of the secant method changed if there are errors in the evaluation of f?

5.7.6. Prove that the secant method converges *r-superlinearly* with *r-order* $(\sqrt{5} + 1)/2$. This is easy.

5.7.7. Show that the secant method converges *q-superlinearly* with *q-order* $(\sqrt{5} + 1)/2$.

5.7.8. The bisection method produces a sequence of intervals (x_l^k, x_r^k) that contain a root of a function of one variable. Given (x_l^k, x_r^k) with $f(x_l^k)f(x_r^k) < 0$ we replace one of the endpoints with $y = (x_l^k + x_r^k)/2$ to force $f(x_l^{k+1})f(x_r^{k+1}) < 0$. If $f(y) = 0$, we terminate the iteration. Show that the sequence $x_k = (x_l^k + x_r^k)/2$ is r-linearly convergent if f is continuous and there exists (x_l^0, x_r^0) such that $f(x_l^0)f(x_r^0) < 0$.

5.7.9. Write a program that solves single nonlinear equations with Newton's method, the chord method, and the secant method. For the secant method, use $x_{-1} = .99x_0$. Apply your program to the following function/initial iterate combinations, tabulate or plot iteration statistics, and explain your results.

1. $f(x) = \cos(x) - x$, $x_0 = .5$
2. $f(x) = \tan^{-1}(x)$, $x_0 = 1$
3. $f(x) = \sin(x)$, $x_0 = 3$
4. $f(x) = x^2$, $x_0 = .5$
5. $f(x) = x^2 + 1$, $x_0 = 10$

5.7.10. For scalar functions of one variable a quadratic model of f about a point x_c is

$$m_c(x) = f(x_c) + f'(x_c)(x - x_c) + f''(x_c)(x - x_c)^2/2.$$

Consider the iterative method that defines x_+ as a root of m_c. What problems might arise in such a method? Resolve these problems and implement your algorithm. Test it on the problems in Exercise 5.7.9. In what cases is this algorithm superior to Newton's method? For hints and generalizations of this idea to functions of several variables, see [170]. Can the ideas in [170] be applied to some of the high-order methods discussed in [190] for equations in one unknown?

5.7.11. Show that if f is a real function of one real variable, f'' is Lipschitz continuous, and $f(x^*) = f'(x^*) = 0$ but $f''(x^*) \neq 0$ then the iteration

$$x_{n+1} = x_n - 2f(x_n)/f'(x_n)$$

converges locally q-quadratically to x^* provided x_0 is sufficiently near to x^*, but not equal to x^* [171].

5.7.12. Show that if f is a real function of one real variable, the standard assumptions hold, $f''(x^*) = 0$ and $f''(x)$ is Lipschitz continuous, then the Newton iteration converges cubically (q-superlinear with q-order 3).

5.7.13. Show that if f is a real function of one real variable, the standard assumptions hold, x_0 is sufficiently near x^*, and $f(x_0)f''(x_0) > 0$, then the Newton iteration converges monotonically to x^*. What if $f'(x^*) = 0$? How can you extend this result if $f'(x^*) = f''(x^*) = \ldots f^{(k)}(x^*) = 0 \neq f^{(k+1)}(x^*)$? See [131].

5.7.14. How would you estimate the q-order of a q-superlinearly convergent sequence? Rigorously justify your answer.

5.7.15. If $x_n \to x^*$ q-superlinearly with q-order $\alpha > 1$, show that $\log(\|e_n\|)$ is a concave function of n in the sense that

$$\log(\|e_{n+1}\|) - \log(\|e_n\|)$$

is a decreasing function of n for n sufficiently large. Is this still the case if the convergence is q-superlinear but there is no q-order? What if the convergence is q-linear?

5.7.16. Show that the sequence on the right side of (5.20) is r-quadratically (but not q-quadratically) convergent to 0 if $\beta\eta\gamma < 1/2$.

5.7.17. Typically (see [89], [184]) the computed LU factorization of $F'(x)$ is the exact LU factorization of a nearby matrix.

$$LU = F'(x) + E.$$

Use the theory in [89], [184] or another book on numerical linear algebra and Theorem 5.4.1 to describe how this factorization error affects the convergence of the Newton iteration. How does the analysis for the QR factorization differ?

5.7.18. Duplicate the results on the H-equation in § 5.6. Try different values of N and c. How does the performance of the methods differ as N and c change? Compare your results to the tabulated values of the H function in [41] or [14]. Compare execution times in your environment. Tabulate or plot iteration statistics. Use the MATLAB cputime command to see how long the entire iteration took for each N, c combination and the MATLAB flops command to get an idea for the number of floating-point operations required by the various methods.

5.7.19. Solve the H-equation with $c = 1$ and $N = 100$. Explain your results (see [50] and [157]). Does the trick in Exercise 5.7.11 improve the convergence? Try one of the approaches in [51], [49], or [118] for acceleration of the convergence. You might also look at [90], [170], [75], [35] and [155] for more discussion of this. Try the chord method and compare the results to part 4 of Exercise 5.7.9 (see [52]).

5.7.20. Solve the H-equation with $N = 100$, $c = .9$ and $c = .9999$ with fixed-point iteration. Compare timings and flop counts with Newton's method, the chord method, and the algorithm in nsol.

5.7.21. Compute by hand the Jacobian of F, the discretized nonlinearity from the H equation in (5.22). Prove that F' can be computed at a cost of less than twice that of an evaluation of F. Modify nsol to accept this analytic Jacobian rather than evaluate a difference Jacobian. How do the execution times and flop counts change?

5.7.22. Modify dirder by setting the forward difference step $h = 10^{-2}$. How does this change affect the convergence rate observations that you have made?

5.7.23. Add noise to your function evaluation with a random number generator, such as the the MATLAB rand function. If the size of the noise is $O(10^{-4})$ and you make the appropriate adjustments to dirder, how are the convergence rates different? What is an appropriate termination criterion? At what point does the iteration stagnate?

5.7.24. Assume that the standard assumptions hold, that the cost of a function evaluation is $O(N^2)$ floating-point operations, the cost of a Jacobian is $O(N)$ function evaluations, and that x_0 is near enough to x^* so that the Newton iteration converges q-quadratically to x^*. Show that the number n of Newton iterations needed to obtain $\|e_n\| \le \epsilon \|e_0\|$ is $O(\log(\epsilon))$ and that the number of floating-point operations required is $O(N^3 \log(\epsilon))$. · What about the chord method?

5.7.25. Develop a sparse version of nsol for problems with tridiagonal Jacobian. In MATLAB, for example, you could modify diffjac to use the techniques of [47] and nsol to use the sparse matrix factorization in MATLAB. Apply your code to the central difference discretization of the two-point boundary value problem

$$-u'' = \sin(u) + f(x), u(0) = u(1) = 0.$$

Here, the solution is $u^*(x) = x(1-x)$, and

$$f(x) = 2 - \sin(x(1-x)).$$

Let $u_0 = 0$ be the initial iterate.

5.7.26. Suppose you try to find an eigenvector-eigenvalue pair (ϕ, λ) of an $N \times N$ matrix A by solving the system of $N + 1$ *nonlinear* equations

(5.23) $$F(\phi, \lambda) = \begin{pmatrix} A\phi - \lambda\phi \\ \phi^T\phi - 1 \end{pmatrix} = \begin{pmatrix} 0 \\ 0 \end{pmatrix}$$

for the vector $x = (\phi^T, \lambda)^T \in R^{N+1}$. What is the Jacobian of this system? If $x = (\phi, \lambda) \in R^{N+1}$ is an eigenvector-eigenvalue pair, when is $F'(x)$ nonsingular? Relate the application of Newton's method to (5.23) to the inverse power method. See [150] for more development of this idea.

Chapter 6

Inexact Newton Methods

Theorem 5.4.1 describes how errors in the derivative/function affect the progress in the Newton iteration. Another way to look at this is to ask how an approximate solution of the linear equation for the Newton step affects the iteration. This was the view taken in [55] where *inexact Newton methods* in which the step satisfies

$$(6.1) \qquad \|F'(x_c)s + F(x_c)\| \le \eta_c \|F(x_c)\|$$

are considered. Any approximate step is accepted provided that the relative residual of the linear equation is small. This is quite useful because conditions like (6.1) are precisely the small linear residual termination conditions for iterative solution of the linear system for the Newton step. Such methods are not new. See [145] and [175], for example, for discussion of these ideas in the context of the classical stationary iterative methods. In most of this chapter we will focus on the approximate solution of the equation for the Newton step by GMRES, but the other Krylov subspace methods discussed in Chapters 2 and 3 and elsewhere can also be used. We follow [69] and refer to the term η_c on the right hand side of (6.1) as the *forcing term*.

6.1. The basic estimates

We will discuss specific implementation issues in § 6.2. Before that we will give the basic result from [55] to show how a step that satisfies (6.1) affects the convergence. We will present this result in two parts. In § 6.1.1 we give a straightforward analysis that uses the techniques of Chapter 5. In § 6.1.2 we show how the requirements of the simple analysis can be changed to admit a more aggressive (i.e., larger) choice of the parameter η.

6.1.1. Direct analysis.
The proof and application of Theorem 6.1.1 should be compared to that of Theorem 5.1.1 and the other results in Chapter 5.

THEOREM 6.1.1. *Let the standard assumptions hold. Then there are δ and K_I such that if $x_c \in \mathcal{B}(\delta)$, s satisfies (6.1), and $x_+ = x_c + s$ then*

$$(6.2) \qquad \|e_+\| \le K_I(\|e_c\| + \eta_c)\|e_c\|.$$

Proof. Let δ be small enough so that the conclusions of Lemma 4.3.1 and Theorem 5.1.1 hold. To prove the first assertion (6.2) note that if[2]

$$r = -F'(x_c)s - F(x_c)$$

is the linear residual then

$$s + F'(x_c)^{-1}F(x_c) = -F'(x_c)^{-1}r$$

and

(6.3) $$e_+ = e_c + s = e_c - F'(x_c)^{-1}F(x_c) - F'(x_c)^{-1}r.$$

Now, (6.1), (4.7), and (4.6) imply that

$$\|s + F'(x_c)^{-1}F(x_c)\| \le \|F'(x_c)^{-1}\|\eta_c\|F(x_c)\|$$

$$\le 4\kappa(F'(x^*))\eta_c\|e_c\|.$$

Hence, using (6.3) and Theorem 5.1.1

$$\|e_+\| \le \|e_c - F'(x_c)^{-1}F(x_c)\| + 4\kappa(F'(x^*))\eta_c\|e_c\|$$

$$\le K\|e_c\|^2 + 4\kappa(F'(x^*))\eta_c\|e_c\|,$$

where K is the constant from (5.2). If we set

$$K_I = K + 4\kappa(F'(x^*)),$$

the proof is complete. □

We summarize the implications of Theorem 6.1.1 for iterative methods in the next result, which is a direct consequence of (6.2) and will suffice to explain most computational observations.

THEOREM 6.1.2. *Let the standard assumptions hold. Then there are δ and $\bar{\eta}$ such that if $x_0 \in \mathcal{B}(\delta)$, $\{\eta_n\} \subset [0, \bar{\eta}]$, then the inexact Newton iteration*

$$x_{n+1} = x_n + s_n,$$

where

$$\|F'(x_n)s_n + F(x_n)\| \le \eta_n\|F(x_n)\|$$

converges q-linearly to x^*. Moreover
- *if $\eta_n \to 0$ the convergence is q-superlinear, and*

- *if $\eta_n \le K_\eta\|F(x_n)\|^p$ for some $K_\eta > 0$ the convergence is q-superlinear with q-order $1 + p$.*

---

[2]Our definition of r is consistent with the idea that the residual in a linear equation is $b - Ax$. In [55] $r = F'(x_c)s + F(x_c)$.

Proof. Let δ be small enough so that (6.2) holds for $x_c \in \mathcal{B}(\delta)$. Reduce δ and $\bar{\eta}$ if needed so that

$$K_I(\delta + \bar{\eta}) < 1,$$

where K_I is from (6.2). Then if $n \geq 0$ and $x_n \in \mathcal{B}(\delta)$ we have

$$\|e_{n+1}\| \leq K_I(\|e_n\| + \eta_n)\|e_n\| \leq K_I(\delta + \bar{\eta})\|e_n\| < \|e_n\|.$$

This proves q-linear convergence with a q-factor of $K_I(\delta + \bar{\eta})$.

If $\eta_n \to 0$ then q-superlinear convergence follows from the definition. If

$$\eta_n \leq K_\eta \|F(x_n)\|^p$$

then we may use (4.7) and (6.2) to conclude

$$\|e_{n+1}\| \leq K_I(\|e_n\|^{1-p} + K_\eta 2\|F'(x^*)\|)\|e_n\|^{1+p}$$

which completes the proof. $\square$

6.1.2. Weighted norm analysis.

Since $K_I = O(\kappa(F'(x^*)))$, one might conclude from Theorem 6.1.2 that if $F'(x^*)$ is ill conditioned very small forcing terms must be used. This is not the case and the purpose of this subsection is to describe more accurately the necessary restrictions on the forcing terms. The results here differ from those in § 6.1.1 in that no restriction is put on the sequence $\{\eta_n\} \subset [0,1)$ other than requiring that 1 not be an accumulation point.

THEOREM 6.1.3. *Let the standard assumptions hold. Then there is δ such that if $x_c \in \mathcal{B}(\delta)$, s satisfies (6.1), $x_+ = x_c + s$, and $\eta_c \leq \eta < \bar{\eta} < 1$, then*

(6.4) $$\|F'(x^*)e_+\| \leq \bar{\eta}\|F'(x^*)e_c\|.$$

Proof. To prove (6.4) note that Theorem 4.0.1 implies that

$$\|F(x_c)\| \leq \|F'(x^*)e_c\| + \frac{\gamma\|e_c\|^2}{2}.$$

Since

$$\|e_c\| = \|F'(x^*)^{-1}F'(x^*)e_c\| \leq \|F'(x^*)^{-1}\|\|F'(x^*)e_c\|$$

we have, with

$$M_0 = \frac{\gamma\|F'(x^*)^{-1}\|}{2}$$

(6.5) $$\|F(x_c)\| \leq (1 + M_0\delta)\|F'(x^*)e_c\|.$$

Now,

$$F'(x^*)e_+ = F'(x^*)(e_c + s)$$

$$= F'(x^*)(e_c - F'(x_c)^{-1}F(x_c) - F'(x_c)^{-1}r).$$

By Theorem 5.1.1

$$\|F'(x^*)(e_c - F'(x_c)^{-1}F(x_c))\| \le K\|F'(x^*)\|\|e_c\|^2.$$

Hence,

(6.6) $$\|F'(x^*)e_+\| \le \|F'(x^*)F'(x_c)^{-1}r\| + K\|F'(x^*)\|\|e_c\|^2.$$

Since

$$\|F'(x^*)F'(x_c)^{-1}r\| \le \|r\| + \|(F'(x^*) - F'(x_c))F'(x_c)^{-1}r\|$$

$$\le (1 + 2\gamma\|F'(x^*)^{-1}\|\|e_c\|)\|r\|,$$

we may set

$$M_1 = 2\gamma\|F'(x^*)^{-1}\|, \text{ and } M_2 = K\|F'(x^*)\|$$

and obtain, using (4.7) and (6.5),

$$\|F'(x^*)e_+\| \le (1 + M_1\delta)\|r\| + M_2\delta\|e_c\|$$

$$\le (1 + M_1\delta)(1 + M_0\delta)\eta_c\|F'(x^*)e_c\|$$

(6.7)
$$+ M_2\delta\|F'(x^*)^{-1}\|\|F'(x^*)e_c\|$$

$$\le ((1 + M_1\delta)(1 + M_0\delta)\eta + M_2\delta\|F'(x^*)^{-1}\|)\|F'(x^*)e_c\|.$$

Now let δ be small enough so that

$$(1 + M_1\delta)(1 + M_0\delta)\eta + M_2\delta\|F'(x^*)^{-1}\| \le \bar{\eta}$$

and the proof is complete. □

Note that the distance δ from the initial iterate to the solution may have to be smaller for (6.4) to hold than for (6.2). However (6.4) is remarkable in its assertion that any method that produces a step with a linear residual less than that of the zero step will reduce the norm of the error if the error is measured with the *weighted norm*

$$\|\cdot\|_* = \|F'(x^*) \cdot \|.$$

In fact, Theorem 6.1.3 asserts q-linear convergence in the weighted norm if the initial iterate is sufficiently near x^*. This is made precise in Theorem 6.1.4.

The application of Theorem 6.1.3 described in Theorem 6.1.4 differs from Theorem 6.1.2 in that we do not demand that $\{\eta_n\}$ be bounded away from 1 by a sufficiently large amount, just that 1 not be an accumulation point. The importance of this theorem for implementation is that a choices of the sequence of forcing terms $\{\eta_n\}$ (such as $\eta_n = .5$ for all n) that try to minimize the number of inner iterations are completely justified by this result if the initial iterate is sufficiently near the solution. We make such a choice in one of the examples considered in § 6.4 and compare it to a modification of a choice from [69] that decreases η_n rapidly with a view toward minimizing the number of outer iterations.

THEOREM 6.1.4. *Let the standard assumptions hold. Then there is δ such that if $x_0 \in \mathcal{B}(\delta)$, $\{\eta_n\} \subset [0, \eta]$ with $\eta < \bar{\eta} < 1$, then the inexact Newton iteration*

$$x_{n+1} = x_n + s_n,$$

where

$$\|F'(x_n)s_n + F(x_n)\| \leq \eta_n \|F(x_n)\|$$

converges q-linearly with respect to $\|\cdot\|_*$ to x^*. Moreover
- *if $\eta_n \to 0$ the convergence is q-superlinear, and*

- *if $\eta_n \leq K_\eta \|F(x_n)\|^p$ for some $K_\eta > 0$ the convergence is q-superlinear with q-order $1 + p$.*

Proof. The proof uses both (6.4) and (6.2). Our first task is to relate the norms $\|\cdot\|$ and $\|\cdot\|_*$ so that we can estimate δ. Let δ_0 be such that (6.4) holds for $\|e_c\| < \delta_0$.

For all $x \in R^N$,

(6.8) $$\|x\|_* \leq \|F'(x^*)\| \|x\| \leq \kappa(F'(x^*)) \|x\|_*,$$

Note that $\|e\| < \delta_0$ if

$$\|e\|_* < \delta_* = \|F'(x^*)\| \delta_0$$

Set $\delta = \|F'(x^*)\|^{-1} \delta_*$. Then $\|e_0\|_* < \delta_*$ if $\|e_0\| < \delta$.

Our proof does not rely on the (possibly false) assertions that $\|e_n\| < \delta$ for all n or that $x_n \to x^*$ q-linearly with respect to the unweighted norm. Rather we note that (6.4) implies that if $\|e_n\|_* < \delta_*$ then

(6.9) $$\|e_{n+1}\|_* \leq \bar{\eta} \|e_n\|_* < \delta_*$$

and hence $x_n \to x^*$ q-linearly with respect to the weighted norm. This proves the first assertion.

To prove the assertions on q-superlinear convergence, note that since $x_n \to x^*$, eventually $\|e_n\|$ will be small enough so that the conclusions of Theorem 6.1.1 hold. The superlinear convergence results then follow exactly as in the proof of Theorem 6.1.2. $\square$

We close this section with some remarks on the two types of results in Theorems 6.1.2 and 6.1.4. The most significant difference is the norm in which q-linear convergence takes place. q-linear convergence with respect to the weighted norm $\|\cdot\|_*$ is equivalent to q-linear convergence of the sequence of nonlinear residuals $\{\|F(x_n)\|\}$. We state this result as Proposition 6.1.1 and leave the proof to the reader in Exercise 6.5.2. The implications for the choice of the forcing terms $\{\eta_n\}$ are also different. While Theorem 6.1.2 might lead one to believe that a small value of η_n is necessary for convergence, Theorem 6.1.4 shows that the sequence of forcing terms need only be kept bounded away from

1. A constant sequence such as $\eta_n = .5$ may well suffice for linear convergence, but at a price of a greater number of outer iterations. We will discuss other factors in the choice of forcing terms in § 6.3.

PROPOSITION 6.1.1. *Let the standard assumptions hold and let* $x_n \to x^*$. *Then* $\|F(x_n)\|$ *converges* q-*linearly to 0 if and only if* $\|e_n\|_*$ *does.*

6.1.3. Errors in the function.

In this section, we state two theorems on inexact local improvement. We leave the proofs, which are direct analogs to the proofs of Theorems 6.1.1 and 6.1.3, to the reader as exercises. Note that errors in the derivative, such as those arising from a difference approximation of the action of $F'(x_c)$ on a vector, can be regarded as part of the inexact solution of the equation for the Newton step. See [55] or Proposition 6.2.1 and its proof for an illustration of this point.

THEOREM 6.1.5. *Let the standard assumptions hold. Then there are* δ *and* K_I *such that if* $x_c \in \mathcal{B}(\delta)$, *s satisfies*

$$(6.10) \qquad \|F'(x_c)s + F(x_c) + \epsilon(x_c)\| \le \eta_c \|F(x_c) + \epsilon(x_c)\|$$

and $x_+ = x_c + s$ *then*

$$(6.11) \qquad \|e_+\| \le K_I((\|e_c\| + \eta_c)\|e_c\| + \|\epsilon(x_c)\|).$$

THEOREM 6.1.6. *Let the standard assumptions hold. Then there is* δ *such that if* $x_c \in \mathcal{B}(\delta)$, *s satisfies* (6.10), $x_+ = x_c + s$, *and* $\eta_c \le \eta < \bar{\eta} < 1$, *then there is* K_E *such that*

$$(6.12) \qquad \|F'(x^*)e_+\| \le \bar{\eta}\|F'(x^*)e_c\| + K_E\|\epsilon(x_c)\|.$$

6.2. Newton-iterative methods

A *Newton-iterative* method realizes (6.1) with an iterative method for the linear system for the Newton step, terminating when the relative linear residual is smaller than η_c (i.e, when (6.1) holds). The name of the method indicates which linear iterative method is used, for example, Newton-SOR, Newton-CG, Newton-GMRES, would use SOR, CG, or GMRES to solve the linear system. This naming convention is taken from [175] and [145]. These methods have also been called *truncated Newton methods* [56], [136] in the context of optimization.

Typically the nonlinear iteration that generates the sequence $\{x_n\}$ is called the *outer iteration* and the linear iteration that generates the approximations to the steps is called the *inner iteration.*

In this section we use the l^2 norm to measure nonlinear residuals. The reason for this is that the Krylov methods use the scalar product and the estimates in Chapters 2 and 3 are in terms of the Euclidean norm. In the examples discussed in § 6.4 we will scale the l^2 norm by a factor of $1/N$ so that the results for the differential and integral equations will be independent of the computational mesh.

6.2.1. Newton GMRES. We provide a detailed analysis of the Newton-GMRES iteration. We begin by discussing the effects of a forward difference approximation to the action of the Jacobian on a vector.

If the linear iterative method is any one of the Krylov subspace methods discussed in Chapters 2 and 3 then each inner iteration requires at least one evaluation of the action of $F'(x_c)$ on a vector. In the case of CGNR and CGNE, an evaluation of the action of $F'(x_c)^T$ on a vector is also required. In many implementations [20], [24], the action of $F'(x_c)$ on a vector w is approximated by a forward difference, (5.15), $D_h F(x:w)$ for some h. It is important to note that this is entirely different from forming the finite difference Jacobian $\nabla_h F(x)$ and applying that matrix to w. In fact, as pointed out in [20], application of GMRES to the linear equation for the Newton step with matrix-vector products approximated by finite differences is the same as the application of GMRES to the matrix $G_h F(x)$ whose last $k-1$ columns are the vectors

$$v_k = D_h F(x:v_{k-1})$$

The sequence $\{v_k\}$ is formed in the course of the forward-difference GMRES iteration.

To illustrate this point, we give the forward-difference GMRES algorithm for computation of the Newton step, $s = -F'(x)^{-1}F(x)$. Note that matrix-vector products are replaced by forward difference approximations to $F'(x)$, a sequence of approximate steps $\{s_k\}$ is produced, that $b = -F(x)$, and that the initial iterate for the linear problem is the zero vector. We give a MATLAB implementation of this algorithm in the collection of MATLAB codes.

ALGORITHM 6.2.1. `fdgmres(s, x, F, h, η, kmax, ρ)`

1. $s = 0$, $r = -F(x)$, $v_1 = r/\|r\|_2$, $\rho = \|r\|_2$, $\beta = \rho$, $k = 0$

2. While $\rho > \eta\|F(x)\|_2$ and $k < kmax$ do

    (a) $k = k + 1$

    (b) $v_{k+1} = D_h F(x:v_k)$
       for $j = 1, \ldots k$

       i. $h_{jk} = v_{k+1}^T v_j$

       ii. $v_{k+1} = v_{k+1} - h_{jk}v_j$

    (c) $h_{k+1,k} = \|v_{k+1}\|_2$

    (d) $v_{k+1} = v_{k+1}/\|v_{k+1}\|_2$

    (e) $e_1 = (1, 0, \ldots, 0)^T \in R^{k+1}$
       Minimize $\|\beta e_1 - H_k y^k\|_{R^{k+1}}$ to obtain $y^k \in R^k$.

    (f) $\rho = \|\beta e_1 - H_k y^k\|_{R^{k+1}}$.

3. $s = V_k y^k$.

In our MATLAB implementation we solve the least squares problem in step 2e by the Givens rotation approach used in Algorithm `gmres`.

As should be clear from inspection of Algorithm gmres, the difference between $\nabla_h F$ and $G_h F$ is that the columns are computed by taking directional derivatives based on two different orthonormal bases: $\nabla_h F$ using the basis of unit vectors in coordinate directions and $G_h F$ the basis for the Krylov space $\mathcal{K}_k$ constructed by algorithm fdgmres. The action of $G_h F$ on $\mathcal{K}_k^\perp$, the orthogonal complement of $\mathcal{K}_k$, is not used by the algorithm and, for the purposes of analysis, could be specified by the action of $\nabla_h F$ on any basis for $\mathcal{K}_k^\perp$. Hence $G_h F$ is also first order accurate in h. Assuming that there is no error in the evaluation of F we may summarize the observations above in the following proposition, which is a special case of the more general results in [20].

PROPOSITION 6.2.1. *Let the standard assumptions hold. Let $\eta \in (0,1)$. Then there are C_G, $\bar{h}$, and δ such that if $x \in \mathcal{B}(\delta)$, $h \leq \bar{h}$, and Algorithm* fdgmres *terminates with $k < kmax$ then the computed step s satisfies*

$$(6.13) \qquad \|F'(x)s + F(x)\|_2 < (\eta + C_G h)\|F(x)\|_2.$$

Proof. First let δ be small enough so that $\mathcal{B}(\delta) \subset \Omega$ and the conclusions to Lemma 4.3.1 hold. Let $\{u_j\}$ be any orthonormal basis for R^N such that $u_j = v_j$ for $1 \leq j \leq k$, where $\{v_j\}_{j=1}^k$ is the orthonormal basis for $\mathcal{K}_k$ generated by Algorithm fdgmres.

Consider the linear map defined by its action on $\{u_j\}$

$$Bu_j = D_h F(x : u_j).$$

Note that

$$D_h F(x : u_j) \;\; = \int_0^1 F'(x + th\|x\|_2 u_j)u_j \, dt$$

$$= F'(x)u_j + \int_0^1 (F'(x + th\|x\|_2 u_j) - F'(x))u_j \, dt.$$

Since F' is Lipschitz continuous and $\{u_j\}$ is an orthonormal basis for R^N we have, with $\bar{\gamma} = \gamma(\|x^*\|_2 + \delta)$,

$$(6.14) \qquad \|B - F'(x)\|_2 \leq h\|x\|_2 \gamma/2 \leq h\bar{\gamma}/2$$

Since the linear iteration (fdgmres) terminates in $k < kmax$ iterations, we have, since B and $G_h F$ agree on $\mathcal{K}_k$,

$$\|Bs + F(x)\|_2 \leq \eta\|F(x)\|_2$$

and therefore

$$(6.15) \qquad \|F'(x)s + F(x)\|_2 \leq \eta\|F(x)\|_2 + h\bar{\gamma}\|s\|_2/2.$$

Assume that

$$\bar{h}\bar{\gamma} \leq \|F'(x^*)^{-1}\|_2^{-1}/2$$

Then Lemma 4.3.1 and (6.15) imply

$$\|F'(x^*)^{-1}\|_2^{-1}\|s\|_2/2 \;\leq\; \|F'(x)^{-1}\|_2^{-1}\|s\|_2$$

$$\leq \|F'(x)s\|_2 \leq (1+\eta)\|F(x)\|_2 + \bar{h}\bar{\gamma}\|s\|_2/2.$$

Therefore,

(6.16) $$\qquad \|s\|_2 \leq 4(1+\eta)\|F'(x^*)^{-1}\|_2\|F(x)\|_2.$$

Combining (6.16) with (6.15) completes the proof with $C_G = 4\bar{\gamma}(1+\eta)\|F'(x^*)^{-1}\|_2$. $\square$

Proposition 6.2.1 implies that a finite difference implementation will not affect the performance of Newton-GMRES if the steps in the forward difference approximation of the derivatives are sufficiently small. In an implementation, therefore, we must specify not only the sequence $\{\eta_n\}$ but also the steps $\{h_n\}$ used to compute forward differences. Note also that Proposition 6.2.1 applies to a restarted GMRES because upon successful termination (6.15) will hold.

We summarize our results so far in the analogs of Theorem 6.1.4 and Theorem 6.1.2. The proofs are immediate consequences of Theorems 6.1.2, 6.1.4, and Proposition 6.2.1 and are left as (easy) exercises.

THEOREM 6.2.1. *Assume that the assumptions of Proposition 6.2.1 hold. Then there are $\delta, \bar{\sigma}$ such that if $x_0 \in \mathcal{B}(\delta)$ and the sequences $\{\eta_n\}$ and $\{h_n\}$ satisfy*

$$\sigma_n = \eta_n + C_G h_n \leq \bar{\sigma}$$

then the forward difference Newton-GMRES iteration

$$x_{n+1} = x_n + s_n$$

where s_n is computed by Algorithm fdgmres *with arguments*

$$(s_n, x_n, F, h_n, \eta_n, kmax, \rho)$$

converges q-linearly and s_n satisfies

(6.17) $$\qquad \|F'(x_n)s_n + F(x_n)\|_2 < \sigma_n\|F(x_n)\|_2.$$

Moreover,

- *if $\sigma_n \to 0$ the convergence is q-superlinear, and*

- *if $\sigma_n \leq K_\eta\|F(x_n)\|_2^p$ for some $K_\eta > 0$ the convergence is q-superlinear with q-order $1 + p$.*

THEOREM 6.2.2. *Assume that the assumptions of Proposition 6.2.1 hold. Then there is δ such that if $x_0 \in \mathcal{B}(\delta)$ and the sequences $\{\eta_n\}$ and $\{h_n\}$ satisfy*

$$\sigma_n = \eta_n + C_G h_n \in [0, \bar{\sigma}]$$

with $0 < \bar{\sigma} < 1$ then the forward difference Newton-GMRES iteration

$$x_{n+1} = x_n + s_n$$

where s_n is computed by Algorithm `fdgmres` *with arguments*

$$(s_n, x_n, F, h_n, \eta_n, kmax, \rho)$$

converges q-linearly with respect to $\| \cdot \|_*$ *and* s_n *satisfies* (6.17). *Moreover*

- *if* $\sigma_n \to 0$ *the convergence is q-superlinear, and*

- *if* $\sigma_n \le K_\eta \|F(x_n)\|_2^p$ *for some* $K_\eta > 0$ *the convergence is q-superlinear with q-order* $1 + p$.

In the case where h_n is approximately the square root of machine roundoff, $\sigma_n \approx \eta_n$ if η_n is not too small and Theorem 6.2.2 states that the observed behavior of the forward difference implementation will be the same as that of an implementation with exact derivatives.

6.2.2. Other Newton-iterative methods.

Other iterative methods may be used as the solver for the linear equation for the Newton step. Newton-multigrid [99] is one approach using a stationary iterative method. See also [6], [111]. If F' is symmetric and positive definite, as it is in the context of unconstrained minimization, [63], Newton-CG is a possible approach.

When storage is restricted or the problem is very large, GMRES may not be practical. GMRES(m), for a small m, may not converge rapidly enough to be useful. CGNR and CGNE offer alternatives, but require evaluation of transpose-vector products. Bi-CGSTAB and TFQMR should be considered in all cases where there is not enough storage for GMRES(m) to perform well.

If a transpose is needed, as it will be if CGNR or CGNE is used as the iterative method, a forward difference formulation is not an option because approximation of the transpose-vector product by differences is not possible. Computation of $\nabla_h F(x)$ and its transpose analytically is one option as is differentiation of F automatically or by hand. The reader may explore this further in Exercise 6.5.13.

The reader may also be interested in experimenting with the other Krylov subspace methods described in § 3.6, [78], and [12].

6.3. Newton-GMRES implementation

We provide a MATLAB code `nsolgm` in the collection that implements a forward-difference Newton-GMRES algorithm. This algorithm requires different inputs than `nsol`. As input we must give the initial iterate, the function, the vector of termination tolerances as we did for `nsol`. In addition, we must provide a method for forming the sequence of forcing terms $\{\eta_n\}$.

We adjust η as the iteration progresses with a variation of a choice from [69]. This issue is independent of the particular linear solver and our discussion is in the general inexact Newton method setting. Setting η to a constant for the entire iteration is often a reasonable strategy as we see in § 6.4, but the choice of that constant depends on the problem. If a constant η is too small, much effort can be wasted in the initial stages of the iteration. The choice in

[69] is motivated by a desire to avoid such *over solving*. Over solving means that the linear equation for the Newton step is solved to a precision far beyond what is needed to correct the nonlinear iteration. As a measure of the degree to which the nonlinear iteration approximates the solution we begin with

$$\eta_n^A = \gamma \|F(x_n)\|^2 / \|F(x_{n-1})\|^2,$$

where $\gamma \in (0,1]$ is a parameter. If η_n^A is uniformly bounded away from 1, then setting $\eta_n = \eta_n^A$ for $n > 0$ would guarantee q-quadratic convergence by Theorem 6.1.1. In this way, the most information possible would be extracted from the inner iteration. In order to specify the choice at $n = 0$ and bound the sequence away from 1 we set

$$(6.18) \qquad \eta_n^B = \begin{cases} \eta_{max}, & n = 0, \\ \min(\eta_{max}, \eta_n^A), & n > 0. \end{cases}$$

In (6.18) the parameter η_{max} is an upper limit on the sequence $\{\eta_n\}$. In [69] the choices $\gamma = .9$ and $\eta_{max} = .9999$ are used.

It may happen that η_n^B is small for one or more iterations while x_n is still far from the solution. A method of *safeguarding* was suggested in [69] to avoid volatile decreases in η_n. The idea is that if η_{n-1} is sufficiently large we do not let η_n decrease by much more than a factor of η_{n-1}.

$$(6.19) \qquad \eta_n^C = \begin{cases} \eta_{max}, & n = 0, \\ \min(\eta_{max}, \eta_n^A), & n > 0, \gamma\eta_{n-1}^2 < .1, \\ \min(\eta_{max}, \max(\eta_n^A, \gamma\eta_{n-1}^2)), & n > 0, \gamma\eta_{n-1}^2 > .1. \end{cases}$$

The constant .1 is somewhat arbitrary. This safeguarding does improve the performance of the iteration.

There is a chance that the final iterate will reduce $\|F\|$ far beyond the desired level and that the cost of the solution of the linear equation for the last step will be more accurate than is really needed. This oversolving on the final step can be controlled comparing the norm of the current nonlinear residual $\|F(x_n)\|$ to the nonlinear residual norm at which the iteration would terminate

$$\tau_t = \tau_a + \tau_r \|F(x_0)\|$$

and bounding η_n from below by a constant multiple of $\tau_t / \|F(x_n)\|$. The algorithm nsolgm does this and uses

$$(6.20) \qquad \eta_n = \min(\eta_{max}, \max(\eta_n^C, .5\tau_t / \|F(x_n)\|)).$$

Exercise 6.5.9 asks the reader to modify nsolgm so that other choices of the sequence $\{\eta_n\}$ can be used.

We use dirder to approximate directional derivatives and use the default value of $h = 10^{-7}$. In all the examples in this book we use the value $\gamma = .9$ as recommended in [69].

ALGORITHM 6.3.1. $\texttt{nsolgm}(x, F, \tau, \eta)$

1. $r_c = r_0 = \|F(x)\|_2/\sqrt{N}$

2. Do while $\|F(x)\|_2/\sqrt{N} > \tau_r r_0 + \tau_a$

    (a) Select η.

    (b) $\texttt{fdgmres}(s, x, F, \eta)$

    (c) $x = x + s$

    (d) Evaluate $F(x)$

    (e) $r_+ = \|F(x)\|_2/\sqrt{N}, \sigma = r_+/r_c, r_c = r_+$

    (f). If $\|F(x)\|_2 \leq \tau_r r_0 + \tau_a$ exit.

Note that the cost of an outer iterate is one evaluation of F to compute the value at the current iterate and other evaluations of F to compute the forward differences for each inner iterate. Hence if a large value of η can be used, the cost of the entire iteration can be greatly reduced. If a maximum of m GMRES iterations is needed (or GMRES(m) is used as a linear solver) the storage requirements of Algorithm $\texttt{nsolgm}$ are the $m + 5$ vectors x, $F(x)$, $x + hv$, $F(x + hv)$, s, and the Krylov basis $\{v_k\}_{k=1}^m$.

Since GMRES forms the inner iterates and makes termination decisions based on scalar products and l^2 norms, we also terminate the outer iteration on small l^2 norms of the nonlinear residuals. However, because of our applications to differential and integral equations, we scale the l^2 norm of the nonlinear residual by a factor of $1/\sqrt{N}$ so that constant functions will have norms that are independent of the computational mesh.

For large and poorly conditioned problems, GMRES will have to be restarted. We illustrate the consequences of this in § 6.4 and ask the reader to make the necessary modifications to $\texttt{nsolgm}$ in Exercise 6.5.9.

The preconditioners we use in § 6.4 are independent of the outer iteration. Since the Newton steps for $MF(x) = 0$ are the same as those for $F(x) = 0$, it is equivalent to precondition the nonlinear equation before calling $\texttt{nsolgm}$.

6.4. Examples for Newton-GMRES

In this section we consider two examples. We revisit the H-equation and solve a preconditioned nonlinear convection-diffusion equation.

In all the figures we plot the relative nonlinear residual $\|F(x_n)\|_2/\|F(x_0)\|_2$ against the number of function evaluations required by all inner and outer iterations to compute x_n. Counts of function evaluations corresponding to outer iterations are indicated by circles. From these plots one can compare not only the number of outer iterations, but also the total cost. This enables us to directly compare the costs of different strategies for selection of η. Note that if only a single inner iteration is needed, the total cost of the outer iteration will be two function evaluations since $F(x_c)$ will be known. One new function evaluation will be needed to approximate the action of $F'(x_c)$ on a vector and then $F(x_+)$ will be evaluated to test for termination. We also count the evaluation of $F(x_0)$ as an additional cost in the evaluation of x_1, which therefore has a minimum cost of three function evaluations. For each example we compare a constant value of η_n with the choice given in (6.20) and used as the default in nsolgm.

6.4.1. Chandrasekhar H-equation.

We solve the H-equation on a 100-point mesh with $c = .9$ using two schemes for selection of the parameter η. The initial iterate is the function identically one. We set the parameters in (6.20) to $\gamma = .9$ and $\eta_{max} = .25$.

We used $\tau_r = \tau_a = 10^{-6}$ in this example.

In Fig. 6.1 we plot the progress of the iteration using $\eta = .1$ with the solid line and using the sequence given by (6.20) with the dashed line. We set the parameters in (6.20) to $\gamma = .9$ and $\eta_{max} = .25$. This choice of η_{max} was the one that did best overall in our experiments on this problem.

We see that the constant η iteration terminates after 12 function evaluations, 4 outer iterations, and roughly 275 thousand floating-point operations. This is a slightly higher overall cost than the other approach in which $\{\eta_n\}$ is given by (6.20), which terminated after 10 function evaluations, 3 outer iterations, and 230 thousand floating-point operations. The chord method with the different (but very similar for this problem) l^∞ termination condition for the same problem, reported in § 5.6 incurred a cost of 3.3 million floating-point operations, slower by a factor of over 1000. This cost is entirely a result of the computation and factorization of a single Jacobian.

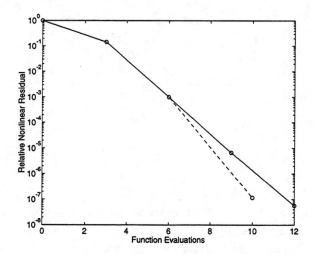

FIG. 6.1. *Newton-GMRES for the* H-*equation, c* = .9.

In Fig. 6.2 the results change for the more ill-conditioned problem with $c = .9999$. Here the iteration with $\eta_n = .1$ performed slightly better and

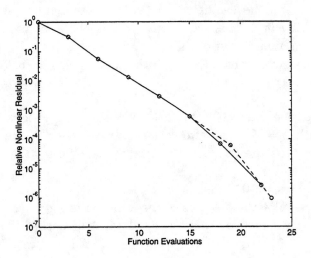

FIG. 6.2.   *Newton-GMRES for the* H-*equation, c = .9999*

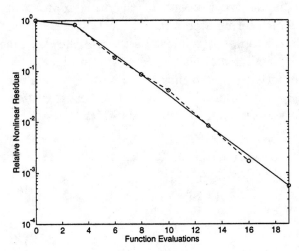

FIG. 6.3.   *Newton-GMRES for the PDE, C = 20.*

terminated after 22 function evaluations and 7 outer iterations, and roughly 513 thousand floating-point operations. The iteration with the decreasing $\{\eta_n\}$ given by (6.20) required 23 function evaluations, 7 outer iterations, and 537 thousand floating-point operations.

6.4.2. Convection-diffusion equation.   We consider the partial differential equation

$$(6.21) \qquad\qquad -\nabla^2 u + Cu(u_x + u_y) = f$$

with homogeneous Dirichlet boundary conditions on the unit square $(0, 1) \times (0, 1)$. f has been constructed so that the exact solution was the discretization of

$$10xy(1 - x)(1 - y) \exp(x^{4.5}).$$

We set $C = 20$ and $u_0 = 0$. As in § 3.7 we discretized (6.21) on a 31×31 grid using centered differences. To be consistent with the second-order accuracy of the difference scheme we used $\tau_r = \tau_a = h^2$, where $h = 1/32$.

We compare the same possibilities for $\{\eta_n\}$ as in the previous example and report the results in Fig. 6.3. In (6.20) we set $\gamma = .9$ and $\eta_{max} = .5$. The computation with constant η terminated after 19 function evaluations and 4 outer iterates at a cost of roughly 3 million floating-point operations. The iteration with $\{\eta_n\}$ given by (6.20) required 16 function evaluations, 4 outer iterations, and 2.6 million floating-point operations. In the computation we preconditioned (6.21) with the fast Poisson solver fish2d.

In this case, the choice of $\{\eta_n\}$ given by (6.20) reduced the number of inner iterations in the early stages of the iteration and avoided oversolving. In cases where the initial iterate is very good and only a single Newton iterate might be needed, however, a large choice of η_{max} or constant η may be the more efficient choice. Examples of such cases include (1) implicit time integration of nonlinear parabolic partial differential equations or ordinary differential equations in which the initial iterate is either the converged solution from the previous time step or the output of a predictor and (2) solution of problems in which the initial iterate is an interpolant of the solution from a coarser mesh.

Preconditioning is crucial for good performance. When preconditioning was not done, the iteration for $C = 20$ required more than 80 function evaluations to converge. In Exercise 6.5.9 the reader is asked to explore this.

6.5. Exercises on inexact Newton methods

6.5.1. Verify (6.8).

6.5.2. Prove Proposition 6.1.1 by showing that if the standard assumptions hold, $0 < \epsilon < 1$, and x is sufficiently near x^* then

$$\|F(x)\|/(1+\epsilon) \le \|F'(x^*)e\| \le (1+\epsilon)\|F(x)\|.$$

6.5.3. Verify (6.14).

6.5.4. Prove Theorems 6.2.2 and 6.2.1.

6.5.5. Prove Theorems 6.1.5 and 6.1.6.

6.5.6. Can anything like Proposition 6.2.1 be proved for a finite-difference Bi-CGSTAB or TFQMR linear solver? If not, what are some possible implications?

6.5.7. Give an example in which $F(x_n) \to 0$ q-linearly but x_n does not converge to x^* q-linearly.

6.5.8. In the DASPK code for integration of differential algebraic equations, [23], preconditioned Newton-GMRES is implemented in such a way that the data needed for the preconditioner is only computed when needed, which may be less often that with each outer iteration. Discuss this strategy and when it might be useful. Is there a relation to the Shamanskii method?

6.5.9. Duplicate the results in § 6.4. For the H-equation experiment with various values of c, such as $c = .5, .99999, .999999$, and for the convection-diffusion equation various values of C such as $C = 1, 10, 40$ and how preconditioning affects the iteration. Experiment with the sequence of forcing terms $\{\eta_n\}$. How do the choices $\eta_n = 1/(n+1)$, $\eta_n = 10^{-4}$ [32], $\eta_n = 2^{1-n}$ [24], $\eta_n = min(\|F(x_n)\|_2, (n+2)^{-1})$ [56], $\eta_n = .05$, and $\eta_n = .75$ affect the results? Present the results of your experiments as graphical iteration histories.

6.5.10. For the H-equation example in § 6.4, vary the parameter c and see how the performance of Newton-GMRES with the various choices of $\{\eta_n\}$ is affected. What happens when $c = 1$? See [119] for an explanation of this.

6.5.11. Are the converged solutions for the preconditioned and unpreconditioned convection-diffusion example in § 6.4 equally accurate?

6.5.12. For the convection-diffusion example in § 6.4, how is the performance affected if GMRES(m) is used instead of GMRES?

6.5.13. Apply Newton-CGNR, Newton-Bi-CGSTAB, or Newton-TFQMR to the two examples in § 6.4. How will you deal with the need for a transpose in the case of CGNR? How does the performance (in terms of time, floating-point operations, function evaluations) compare to Newton-GMRES? Experiment with the forcing terms. The MATLAB codes `fdkrylov`, which allows you to choose from a variety of forward difference Krylov methods, and `nsola`, which we describe more fully in Chapter 8, might be of use in this exercise.

6.5.14. If you have done Exercise 5.7.25, apply `nsolgm` to the same two-point boundary value problem and compare the performance.

Chapter 7

Broyden's method

Quasi-Newton methods maintain approximations of the solution x^* and the Jacobian at the solution $F'(x^*)$ as the iteration progresses. If x_c and B_c are the current approximate solution and Jacobian, then

$$(7.1) \qquad x_+ = x_c - B_c^{-1} F(x_c).$$

After the computation of x_+, B_c is *updated* to form B_+. The construction of B_+ determines the quasi-Newton method.

The advantages of such methods, as we shall see in § 7.3, is that solution of equations with the quasi-Newton approximation to the Jacobian is often much cheaper that using $F'(x_n)$ as the coefficient matrix. In fact, if the Jacobian is dense, the cost is little more that that of the chord method. The method we present in this chapter, Broyden's method, is locally superlinearly convergent, and hence is a very powerful alternative to Newton's method or the chord method.

In comparison with Newton-iterative methods, quasi-Newton methods require only one function evaluation for each nonlinear iterate; there is no cost in function evaluations associated with an inner iteration. Hence, if a good preconditioner (initial approximation to $F'(x^*)$) can be found, these methods could have an advantage in terms of function evaluation cost over Newton-iterative methods.

Broyden's method [26] computes B_+ by

$$(7.2) \qquad B_+ = B_c + \frac{(y - B_c s)s^T}{s^T s} = B_c + \frac{F(x_+)s^T}{s^T s}.$$

In (7.2) $y = F(x_+) - F(x_c)$ and $s = x_+ - x_c$.

Broyden's method is an example of a *secant update*. This means that the updated approximation to $F'(x^*)$ satisfies the *secant equation*

$$(7.3) \qquad B_+ s = y.$$

In one dimension, (7.3) uniquely specifies the classical secant method. For equations in several variables (7.3) alone is a system of N equations in N^2

unknowns. The role of the secant equation is very deep. See [62] and [64] for discussion of this topic. In this book we focus on Broyden's method and our methods of analysis are not as general as those in [62] and [64]. In this chapter we will assume that the function values are accurate and refer the reader to [66], [65], and [114] for discussions of quasi-Newton methods with inaccurate data. Locally superlinearly convergent secant methods can be designed to maintain the sparsity pattern, symmetry, or positive definiteness, of the approximate Jacobian, and we refer the reader to [28], [63], [62], and [64] for discussion and analysis of several such methods. More subtle structural properties can also be preserved by secant methods. Some examples can be found in [101], [96], [95], [113], and [116].

Broyden's method is also applicable to linear equations [29], [82], [84], [104], [143], [130], $Ax = b$, where $B \approx A$ is updated. One can also apply the method to linear least squares problems [84], [104]. The analysis is somewhat clearer in the linear case and the plan of this chapter is to present results for the linear case for both the theory and the examples. One interesting result, which we will not discuss further, is that for linear problems Broyden's method converges in $2N$ iterations [29], [82], [84], [143].

We will express our results in terms of the error in the Jacobian

$$(7.4) \qquad\qquad E = B - F'(x^*)$$

and the step $s = x_+ - x_c$. While we could use $E = B - F'(x)$, (7.4) is consistent with our use of $e = x - x^*$. When indexing of the steps is necessary, we will write

$$s_n = x_{n+1} - x_n.$$

In this chapter we show that if the data x_0 and B_0 are sufficiently good, the Broyden iteration will converge q-superlinearly to the root.

7.1. The Dennis–Moré condition

In this section we consider general iterative methods of the form

$$(7.5) \qquad\qquad x_{n+1} = x_n - B_n^{-1}F(x_n)$$

where $B_n = F'(x^*) + E_n \approx F'(x^*)$ is generated by some method (not necessarily Broyden's method).

Verification of q-superlinear convergence cannot be done by the direct approaches used in Chapters 5 and 6. We must relate the superlinear convergence condition that $\eta_n \to 0$ in Theorem 6.1.2 to a more subtle, but easier to verify condition in order to analyze Broyden's method. This technique is based on verification of the *Dennis–Moré condition* [61], [60] on the sequence of steps $\{s_n\}$ and errors in the Jacobian $\{E_n\}$

$$(7.6) \qquad\qquad \lim_{n \to \infty} \frac{\|E_n s_n\|}{\|s_n\|} = 0.$$

The main result in this section, Theorem 7.1.1 and its corollary for linear problems, are used to prove superlinear convergence for Broyden's method in this chapter and many other quasi-Newton methods. See [61], [28], [62], and [64] for several more examples. Our formulation of the Dennis–Moré result is a bit less general than that in [60] or [63].

THEOREM 7.1.1. *Let the standard assumptions hold, let $\{B_n\}$ be a sequence of nonsingular $N \times N$ matrices, let $x_0 \in R^N$ be given and let $\{x_n\}_{n=1}^{\infty}$ be given by (7.5). Assume that $x_n \neq x^*$ for any n. Then $x_n \to x^*$ q-superlinearly if and only if $x_n \to x^*$ and the Dennis–Moré condition (7.6) holds.*

Proof. Since

$$-F(x_n) = B_n s_n = F'(x^*)s_n + E_n s_n$$

we have

$$(7.7) \quad E_n s_n = -F'(x^*)s_n - F(x_n) = -F'(x^*)e_{n+1} + F'(x^*)e_n - F(x_n).$$

We use the fundamental theorem of calculus and the standard assumptions to obtain

$$F'(x^*)e_n - F(x_n) = \int_0^1 (F'(x^*) - F'(x^* + te_n))e_n \, dt$$

and hence

$$\|F'(x^*)e_n - F(x_n)\| \leq \gamma \|e_n\|^2/2.$$

Therefore, by (7.7)

$$(7.8) \qquad \|E_n s_n\| \leq \|F'(x^*)e_{n+1}\| + \gamma \|e_n\|^2/2.$$

Now, if $x_n \to x^*$ q-superlinearly, then for n sufficiently large

$$(7.9) \qquad \|s_n\|/2 \leq \|e_n\| \leq 2\|s_n\|.$$

The assumption that $x_n \neq x^*$ for any n implies that the sequence

$$(7.10) \qquad \nu_n = \|e_{n+1}\|/\|e_n\|$$

is defined and and $\nu_n \to 0$ by superlinear convergence of $\{x_n\}$. By (7.9) we have

$$\|e_{n+1}\| = \nu_n \|e_n\| \leq 2\nu_n \|s_n\|,$$

and hence (7.8) implies that

$$\|E_n s_n\| \leq (2\|F'(x^*)\|\nu_n + \gamma \|e_n\|)\|s_n\|$$

which implies the Dennis-Moré condition (7.6).

Conversely, assume that $x_n \to x^*$ and that (7.6) holds. Let

$$\mu_n = \frac{\|E_n s_n\|}{\|s_n\|}.$$

We have, since

(7.11) $$E_n s_n = (B_n - F'(x^*))s_n = -F(x_n) - F'(x^*)s_n,$$

(7.12) $$\|s_n\|\|F'(x^*)^{-1}\|^{-1} \leq \|F'(x^*)s_n\| \leq \|E_n s_n\| + \|F(x_n)\|$$
$$= \mu_n\|s_n\| + \|F(x_n)\|.$$

Since $\mu_n \to 0$ by assumption,

(7.13) $$\mu_n \leq \|F'(x^*)^{-1}\|^{-1}/2,$$

for n sufficiently large. We assume now that n is large enough so that (7.13) holds. Hence,

(7.14) $$\|s_n\| \leq 2\|F'(x^*)^{-1}\|\|F(x_n)\|.$$

Moreover, using the standard assumptions and (7.11) again,

(7.15) $$\|F(x_n) + F'(x_n)s_n\| \leq \|F(x_n) + F'(x^*)s_n\| + \|(F'(x^*) - F'(x_n))s_n\|$$
$$\leq (\mu_n + \gamma\|e_n\|)\|s_n\|.$$

Since $x_n \to x^*$ by assumption, $\eta_n \to 0$ where

$$\eta_n = 2\|F'(x^*)^{-1}\|(\mu_n + \gamma\|e_n\|).$$

Combining (7.14) and (7.15) implies that

(7.16) $$\|F(x_n) + F'(x_n)s_n\| \leq \eta_n\|F(x_n)\|,$$

which is the inexact Newton condition. Since $\eta_n \to 0$, the proof is complete by Theorem 6.1.2. □

7.2. Convergence analysis

Our convergence analysis is based on an approach to infinite-dimensional problems from [97]. That paper was the basis for [168], [104], and [115], and we employ notation from all four of these references. We will use the l^2 norm throughout our discussion. We begin with a lemma from [104].

LEMMA 7.2.1. *Let* $0 < \hat{\theta} < 1$ *and let*

$$\{\theta_n\}_{n=0}^\infty \subset (\hat{\theta}, 2 - \hat{\theta}).$$

Let $\{\epsilon_n\}_{n=0}^\infty \subset R^N$ *be such that*

$$\sum_n \|\epsilon_n\|_2 < \infty,$$

and let $\{\eta_n\}_{n=0}^\infty$ *be a set of vectors in* R^N *such that* $\|\eta_n\|_2$ *is either 1 or 0 for all* n. *Let* $\psi_0 \in R^N$ *be given. If* $\{\psi_n\}_{n=1}^\infty$ *is given by*

(7.17) $$\psi_{n+1} = \psi_n - \theta_n(\eta_n^T\psi_n)\eta_n + \epsilon_n$$

then

(7.18) $$\lim_{n\to\infty} \eta_n^T\psi_n = 0.$$

Proof. We first consider the case in which $\epsilon_n = 0$ for all n. In that case, we can use the fact that

$$\theta_n(2 - \theta_n) > \hat{\theta}^2 > 0$$

to show that the sequence $\{\psi_n\}$ is bounded in l_2-norm by $\|\psi_0\|_2$ and, in fact,

$$\|\psi_{n+1}\|_2^2 \leq \|\psi_n\|_2^2 - \theta_n(2 - \theta_n)(\eta_n^T \psi_n)^2$$

$$\leq \|\psi_n\|_2^2 - \hat{\theta}^2(\eta_n^T \psi_n)^2$$

$$\leq \|\psi_n\|_2^2.$$

Therefore, for any $M > 0$,

$$\sum_{n=0}^{M} (\eta_n^T \psi_n)^2 \leq \frac{\|\psi_0\|_2^2 - \|\psi_{M+1}\|_2^2}{\hat{\theta}^2} \leq \frac{\|\psi_0\|_2}{\hat{\theta}^2}.$$

We let $M \to \infty$ to obtain

$$\text{(7.19)} \qquad \sum_{n=0}^{\infty} (\eta_n^T \psi_n)^2 < \infty.$$

Convergence of the series in (7.19) implies convergence to zero of the terms in the series, which is (7.18).

To prove the result for $\epsilon_n \neq 0$ we use the inequality

$$\text{(7.20)} \qquad \sqrt{a^2 - b^2} \leq a - \frac{b^2}{2a},$$

which is valid for $a > 0$ and $|b| \leq a$. This inequality is used often in the analysis of quasi-Newton methods [63]. From (7.20) we conclude that if $\psi_n \neq 0$ then

$$\|\psi_n - \theta_n(\eta_n, \psi_n)\eta_n\|_2 \leq \sqrt{\|\psi_n\|_2^2 - \theta_n(2 - \theta_n)(\eta_n^T \psi_n)^2}$$

$$\leq \|\psi_n\|_2 - \frac{\theta_n(2 - \theta_n)(\eta_n^T \psi_n)^2}{2\|\psi_n\|_2}.$$

Hence if $\psi_n \neq 0$

$$\text{(7.21)} \qquad \|\psi_{n+1}\|_2 \leq \|\psi_n\|_2 - \frac{\theta_n(2 - \theta_n)(\eta_n^T \psi_n)^2}{2\|\psi_n\|_2} + \|\epsilon_n\|_2.$$

Hence

$$\text{(7.22)} \qquad (\eta_n^T \psi_n)^2 \leq \frac{2\|\psi_n\|_2}{\theta_n(2 - \theta_n)}(\|\psi_n\|_2 - \|\psi_{n+1}\|_2 + \|\epsilon_n\|_2),$$

which holds even if $\psi_n = 0$.

From (7.21) we conclude that

$$\|\psi_{n+1}\|_2 \leq \mu,$$

where

$$\mu = \sum_{i=0}^{\infty} \|\epsilon_i\|_2 + \|\psi_0\|_2.$$

Hence

$$\sum_{n=0}^{M}(\eta_n^T\psi_n)^2 \leq \frac{2\mu}{\hat{\theta}^2}\sum_{n=0}^{M}(\|\psi_n\|_2 - \|\psi_{n+1}\|_2 + \|\epsilon_n\|_2)$$

$$= \frac{2\mu}{\hat{\theta}^2}\left(\|\psi_0\|_2 - \|\psi_{M+1}\|_2 + \sum_{n=0}^{M}\|\epsilon_n\|_2\right)$$

$$\leq \frac{2\mu^2}{\hat{\theta}^2}.$$

Hence (7.19) holds and the proof is complete. $\square$

7.2.1. Linear problems.   We may use Lemma 7.2.1 to prove a result from [130] on the convergence of Broyden's method for linear problems. The role of the parameter $\hat{\theta}$ and the sequence $\{\theta_n\}$ in Lemma 7.2.1 is nicely illustrated by this application of the lemma.

In this section $F(x) = Ax - b$ and $B_n = A + E_n$. The standard assumptions in this case reduce to nonsingularity of A. Our first task is to show that the errors in the Broyden approximations to A do not grow. This property, called *bounded deterioration* [28], is an important part of the convergence analysis. The reader should compare the statement of the result with that in the nonlinear case.

In the linear case, the Broyden update has a very simple form. We will consider a modified update

$$(7.23)\qquad B_+ = B_c + \theta_c\frac{(y - B_c s)s^T}{s^T s} = B_c + \theta_c\frac{(Ax_+ - b)s^T}{s^T s}.$$

If x_c and B_c are the current approximations to $x^* = A^{-1}b$ and A and $s = x_+ - x_c$ then

$$y - B_c s = (Ax_+ - Ax_c) - B_c(x_+ - x_c) = -E_c s$$

and therefore

$$(7.24)\qquad B_+ = B_c + \theta_c\frac{(y - B_c s)s^T}{\|s\|_2^2} = B_c - \theta_c\frac{(E_c s)s^T}{\|s\|_2^2}.$$

LEMMA 7.2.2. *Let A be nonsingular, $\theta_c \in [0, 2]$, and $x_c \in R^N$ be given. Let B_c be nonsingular and let B_+ be formed by the Broyden update (7.23). Then*

$$(7.25)\qquad \|E_+\|_2 \leq \|E_c\|_2.$$

Proof. By subtracting A from both sides of (7.24) we have

(7.26)
$$E_+ = E_c - \theta_c \frac{(E_c s) s^T}{\|s\|_2^2}$$

$$= E_c(I - \theta_c P_s),$$

where P_s is the orthogonal projector

(7.27)
$$P_s = \frac{s s^T}{\|s\|_2^2}.$$

This completes the proof as

$$\|E_+\|_2 \le \|E_c\|_2 \|I - \theta_c P_s\|_2$$

and $\|I - \theta_c P_s\|_2 \le 1$ by orthogonality of P_s and the fact that $0 \le \theta_c \le 2$. $\square$

Now, θ_c can always be selected to make B_+ nonsingular. In [130] one suggestion was

$$\theta_c = \begin{cases} 1, & |\gamma_c| \ge \sigma, \\ \dfrac{1 - \text{sign}(\gamma_c)\sigma}{1 - \gamma_c}, & \text{otherwise,} \end{cases}$$

where

$$\gamma_c = \frac{(B_c^{-1} y)^T s}{\|s\|_2^2} = \frac{(B_c^{-1} A s)^T s}{\|s\|_2^2}$$

and $\sigma \in (0, 1)$ is fixed. However the results in [130] assume only that the sequence $\{\theta_n\}$ satisfies the hypotheses of Lemma 7.2.1 for some $\hat\theta \in (0, 1)$ and that θ_c is always chosen so that B_+ is nonsingular.

For linear problems one can show that the Dennis–Moré condition implies convergence and hence the assumption that convergence takes place is not needed. Specifically

PROPOSITION 7.2.1. *Let A be nonsingular. Assume that $\{\theta_n\}$ satisfies the hypotheses of Lemma 7.2.1 for some $\hat\theta \in (0, 1)$. If $\{\theta_n\}$ is such that the matrices B_n obtained by (7.23) are nonsingular and $\{x_n\}$ is given by the modified Broyden iteration*

(7.28)
$$x_{n+1} = x_n - B_n^{-1}(A x_n - b)$$

then the Dennis–Moré condition implies convergence of $\{x_n\}$ to $x^* = A^{-1}b$.

We leave the proof as Exercise 7.5.1.

The superlinear convergence result is remarkable in that the initial iterate need not be near the root.

THEOREM 7.2.1. *Let A be nonsingular. Assume that $\{\theta_n\}$ satisfies the hypotheses of Lemma 7.2.1 for some $\hat\theta \in (0, 1)$. If $\{\theta_n\}$ is such the matrices B_n obtained by (7.23) are nonsingular then the modified Broyden iteration (7.28) converges q-superlinearly to $x^* = A^{-1}b$ for every initial iterate x_0.*

Proof. Our approach is to show that for every $\phi \in R^N$ that

$$(7.29) \qquad \lim_{n \to \infty} \phi^T \left(\frac{E_n s_n}{\|s_n\|_2} \right) = 0.$$

Assuming that (7.29) holds, setting $\phi = e_j$, the unit vector in the jth coordinate direction, shows that the j component of $\dfrac{E_n s_n}{\|s_n\|_2}$ has limit zero. As j was arbitrary, we have

$$\lim_{n \to \infty} \frac{E_n s_n}{\|s_n\|_2} = 0$$

which is the Dennis–Moré condition. This will imply q-superlinear convergence by Theorem 7.1.1 and Proposition 7.2.1.

Let $\phi \in R^N$ be given. Let

$$P_n = \frac{s_n s_n^T}{\|s_n\|_2^2}.$$

Since for all $v \in R^N$ and all n

$$v^T (E_n^T \phi) = \phi^T (E_n v)$$

and, by (7.26)

$$E_{n+1}^T \phi = (I - \theta_n P_n) E_n^T \phi,$$

we may invoke Lemma 7.2.1 with

$$\eta_n = s_n / \|s_n\|_2, \ \psi_n = E_n^T \phi, \ \text{and} \ \epsilon_n = 0$$

to conclude that

$$(7.30) \qquad \eta_n^T \psi_n = \frac{(E_n^T \phi)^T s_n}{\|s_n\|_2} = \phi^T \frac{E_n s_n}{\|s_n\|_2} \to 0.$$

This completes the proof. $\square$

The result here is not a local convergence result. It is global in the sense that q-superlinear convergence takes place independently of the initial choices of x and B. It is known [29], [82], [84], [143], that the Broyden iteration will terminate in $2N$ steps in the linear case.

7.2.2. Nonlinear problems.

Our analysis of the nonlinear case is different from the classical work [28], [63]. As indicated in the preface, we provide a proof based on ideas from [97], [115], and [104], that does not use the Frobenius norm and extends to various infinite-dimensional settings.

For nonlinear problems our analysis is local and we set $\theta_n = 1$. However we must show that the sequence $\{B_n\}$ of Broyden updates exists and remains nonsingular. We make a definition that allows us to compactly express this fact.

DEFINITION 7.2.1. *We say that the* Broyden sequence $(\{x_n\}, \{B_n\})$ *for the data* (F, x_0, B_0) *exists if F is defined at x_n for all n, B_n is nonsingular for all n, and x_{n+1} and B_{n+1} can be computed from x_n and B_n using (7.1) and (7.2).*

We will show that if the standard assumptions hold and x_0 and B_0 are sufficiently good approximations to x^* and $F'(x^*)$ then the Broyden sequence exists for the data (F, x_0, B_0) and that the Broyden iterates converge q-superlinearly to x^*. The development of the proof is complicated. We first prove a bounded deterioration result like Lemma 7.2.2. However in the nonlinear case, the errors in the Jacobian approximation can grow as the iteration progresses. We then show that if the initial data is sufficiently good, this growth can be limited and therefore the Broyden sequence exists and the Broyden iterates converges to x^* at least q-linearly. Finally we verify the Dennis–Moré condition (7.6), which, together with the q-linear convergence proved earlier completes the proof of local q-superlinear convergence.

Bounded deterioration. Our first task is to show that bounded deterioration holds. Unlike the linear case, the sequence $\{\|E_n\|\}$ need not be monotone non-increasing. However, the possible increase can be bounded in terms of the errors in the current and new iterates.

THEOREM 7.2.2. *Let the standard assumptions hold. Let $x_c \in \Omega$ and a nonsingular matrix B_c be given. Assume that*

$$x_+ = x_c - B_c^{-1}F(x_c) = x_c + s \in \Omega$$

and B_+ is given by (7.2). Then

$$(7.31) \qquad \|E_+\|_2 \le \|E_c\| + \gamma(\|e_c\|_2 + \|e_+\|_2)/2$$

Proof. Note that (7.7) implies that

$$
\begin{aligned}
E_c s &= -F(x_+) + (F(x_+) - F(x_c) - F'(x^*)s) \\
&= -F(x_+) + \int_0^1 (F'(x_c + ts) - F'(x^*))s\, dt.
\end{aligned}
$$
(7.32)

We begin by writing (7.32) as

$$F(x_+) = -E_c s + \int_0^1 (F'(x_c + ts) - F'(x^*))s\, dt$$

and then using (7.2) to obtain

$$E_+ = E_c(I - P_s) + \frac{(\Delta_c s)s^T}{\|s\|_2^2}$$

where P_s is is given by (7.27) and

$$\Delta_c = \int_0^1 (F'(x_c + ts) - F'(x^*))\, dt.$$

Hence, by the standard assumptions

$$\|\Delta_c\|_2 \leq \gamma(\|e_c\|_2 + \|e_+\|_2)/2.$$

Just as in the proof of Lemma 7.2.2 for the linear case

$$\|E_+\|_2 \leq \|E_c\|_2 + \|\Delta_c\|_2$$

and the proof is complete. $\square$

Local q-linear convergence.   Theorem 7.2.3 states that Broyden's method is locally q-linearly convergent. The proof is more complex than similar results in Chapters 5 and 6. We must explore the relation between the size of $\|E_0\|$ needed to enforce q-linear convergence and the q-factor.

THEOREM 7.2.3. *Let the standard assumptions hold. Let $r \in (0,1)$ be given. Then there are δ and δ_B such that if $x_0 \in \mathcal{B}(\delta)$ and $\|E_0\|_2 < \delta_B$ the Broyden sequence for the data (F, x_0, B_0) exists and $x_n \to x^*$ q-linearly with q-factor at most r.*

Proof. Let δ and δ_1 be small enough so that the conclusions of Theorem 5.4.3 hold. Then, reducing δ and δ_1 further if needed, the q-factor is at most

$$K_A(\delta + \delta_1) \leq r,$$

where K_A is from the statement of Theorem 5.4.3.

We will prove the result by choosing δ_B and reducing δ if necessary so that $\|E_0\|_2 \leq \delta_B$ will imply that $\|E_n\|_2 \leq \delta_1$ for all n, which will prove the result. To do this reduce δ if needed so that

$$(7.33) \qquad \qquad \delta_2 = \frac{\gamma(1+r)\delta}{2(1-r)} < \delta_1$$

and set

$$(7.34) \qquad \qquad \delta_B = \delta_1 - \delta_2.$$

We show that $\|E_n\|_2 \leq \delta_1$ by induction. Since $\|E_0\|_2 < \delta_B < \delta_1$ we may begin the induction. Now assume that $\|E_k\|_2 < \delta_1$ for all $0 \leq k \leq n$. By Theorem 7.2.2,

$$\|E_{n+1}\|_2 \leq \|E_n\|_2 + \gamma(\|e_{n+1}\|_2 + \|e_n\|_2)/2.$$

Since $\|E_n\|_2 < \delta_1$, $\|e_{n+1}\|_2 \leq r\|e_n\|_2$ and therefore

$$\|E_{n+1}\|_2 \leq \|E_n\|_2 + \gamma(1+r)\|e_n\|_2/2 \leq \|E_n\|_2 + \gamma(1+r)r^n\delta/2.$$

We can estimate $\|E_n\|_2, \|E_{n-1}\|_2, \ldots$ in the same way to obtain

$$\|E_{n+1}\|_2 \ \leq \|E_0\|_2 + \frac{(1+r)\gamma\delta}{2} \sum_{j=0}^{n} r^j$$

$$\leq \delta_B + \frac{(1+r)\gamma\delta}{2(1-r)} \leq \delta_1,$$

which completes the proof. $\square$

Verification of the Dennis–Moré condition.   The final task is to verify the Dennis–Moré condition.

THEOREM 7.2.4. *Let the standard assumptions hold. Then there are δ and δ_B such that if $x_0 \in \mathcal{B}(\delta)$ and $\|E_0\|_2 < \delta_B$ the Broyden sequence for the data (F, x_0, B_0) exists and $x_n \to x^*$ q-superlinearly.*

Proof. Let δ and δ_B be such that the conclusions of Theorem 7.2.3 hold. By Theorem 7.1.1 we need only show that Dennis–Moré condition (7.6) holds to complete the proof.

As in the proof of Theorem 7.2.1 let

$$P_n = \frac{s_n s_n^T}{\|s_n\|_2^2}.$$

Set

$$\Delta_n = \int_0^1 (F'(x_n + ts_n) - F'(x^*))\, dt.$$

Let $\phi \in R^N$ be arbitrary. Note that

$$E_{n+1}^T \phi = (I - P_n)E_n^T \phi + P_n \Delta_n^T \phi.$$

We wish to apply Lemma 7.2.1 with

$$\psi_n = E_n^T \phi, \eta_n = s_n/\|s_n\|_2, \text{ and } \epsilon_n = P_n \Delta_n^T \phi.$$

The hypothesis in the lemma that

$$\sum_n \|\epsilon_n\|_2 < \infty$$

holds since Theorem 7.2.3 and the standard assumptions imply

$$\|\Delta_n\|_2 \leq \gamma(1 + r)r^n \delta/2.$$

Hence Lemma 7.2.1 implies that

$$\eta_n^T \psi_n = \frac{(E_n^T \phi)^T s_n}{\|s_n\|_2} = \phi^T \frac{E_n s_n}{\|s_n\|_2} \to 0.$$

This, as in the proof of Theorem 7.2.1, implies the Dennis–Moré condition and completes the proof. □

7.3. Implementation of Broyden's method

The implementation of Broyden's method that we advocate here is related to one that has been used in computational fluid mechanics [73]. Some such methods are called *limited memory* formulations in the optimization literature [138], [142], [31]. In these methods, after the storage available for the iteration is exhausted, the oldest of the stored vectors is replaced by the most recent. Another approach, *restarting*, clears the storage and starts over.

This latter approach is similar to the restarted GMRES iteration. Our basic implementation is a nonlinear version of the implementation for linear problems in [67]. We use a restarted approach, though a limited memory approach could be used as well.

We begin by showing how the initial approximation to $F'(x^*)$, B_0, may be regarded as a preconditioner and incorporated into the defintion of F just as preconditioners were in the implementation of Newton-GMRES in § 6.4.

LEMMA 7.3.1. *Assume that the Broyden sequence* $(\{x_n\}, \{B_n\})$ *for the data* (F, x_0, B_0) *exists. Then the Broyden sequence* $(\{y_n\}, \{C_n\})$ *exists for the data* $(B_0^{-1}F, x_0, I)$ *and*

$$(7.35) \qquad\qquad x_n = y_n, \quad B_n = B_0 C_n$$

for all n.

Proof. We proceed by induction. Equation (7.35) holds by definition when $n = 0$. If (7.35) holds for a given n, then

$$y_{n+1} = y_n - C_n^{-1} B_0^{-1} F(y_n)$$

$$(7.36) \qquad = x_n - (B_0 C_n)^{-1} F(x_n) = x_n - B_n^{-1} F(x_n)$$

$$= x_{n+1}.$$

Now $s_n = x_{n+1} - x_n = y_{n+1} - y_n$ by (7.36). Hence

$$B_{n+1} = B_n + \frac{F(x_{n+1}) s_n^T}{\|s_n\|_2^2}$$

$$= B_0 C_n + B_0 \frac{B_0^{-1} F(y_{n+1}) s_n^T}{\|s_n\|_2^2} = B_0 C_{n+1}.$$

This completes the proof. □

The consequence of this proof for implementation is that we can assume that $B_0 = I$. If a better choice for B_0 exists, we can incorporate that into the function evaluation. Our plan will be to store B_n^{-1} in a compact and easy to apply form. The basis for this is the following simple formula [68], [176], [177], [11].

PROPOSITION 7.3.1. *Let B be a nonsingular $N \times N$ matrix and let $u, v \in R^N$. Then $B + uv^T$ is invertible if and only if $1 + v^T B^{-1} u \neq 0$. In this case,*

$$(7.37) \qquad (B + uv^T)^{-1} = \left(I - \frac{(B^{-1}u)v^T}{1 + v^T B^{-1} u} \right) B^{-1}.$$

The expression (7.37) is called the *Sherman–Morrison formula*. We leave the proof to Exercise 7.5.2.

In the context of a sequence of Broyden updates $\{B_n\}$ we have for $n \geq 0$,

$$B_{n+1} = B_n + u_n v_n^T,$$

where

$$u_n = F(x_{n+1})/\|s_n\|_2 \text{ and } v_n = s_n/\|s_n\|_2.$$

Setting

$$w_n = (B_n^{-1}u_n)/(1 + v_n^T B_n^{-1} u_n)$$

we see that, if $B_0 = I$,

(7.38)
$$\begin{aligned} B_n^{-1} &= (I - w_{n-1}v_{n-1}^T)(I - w_{n-2}v_{n-2}^T)\ldots(I - w_0 v_0^T) \\ &= \prod_{j=0}^{n-1}(I - w_j v_j^T). \end{aligned}$$

Since the empty matrix product is the identity, (7.38) is valid for $n \geq 0$.

Hence the action of B_n^{-1} on $F(x_n)$ (i.e., the computation of the Broyden step) can be computed from the $2n$ vectors $\{w_j, v_j\}_{j=0}^{n-1}$ at a cost of $O(Nn)$ floating-point operations. Moreover, the Broyden step for the following iteration is

(7.39)
$$s_n = -B_n^{-1}F(x_n) = -\prod_{j=0}^{n-1}(I - w_j v_j^T)F(x_n).$$

Since the product

$$\prod_{j=0}^{n-2}(I - w_j v_j^T)F(x_n)$$

must also be computed as part of the computation of w_{n-1} we can combine the computation of w_{n-1} and s_n as follows:

(7.40)
$$\begin{aligned} w &= \prod_{j=0}^{n-2}(I - w_j v_j^T)F(x_n) \\ w_{n-1} &= C_w w \text{ where } C_w = (\|s_{n-1}\|_2 + v_{n-1}^T w)^{-1} \\ s_n &= -(I - w_{n-1}v_{n-1}^T)w. \end{aligned}$$

One may carry this one important step further [67] and eliminate the need to store the sequence $\{w_n\}$. Note that (7.40) implies that for $n \geq 1$

$$\begin{aligned} s_n &= -w + C_w w(v_{n-1}^T w) = w(-1 + C_w(C_w^{-1} - \|s_{n-1}\|_2)) \\ &= -C_w w \|s_{n-1}\|_2 = -\|s_{n-1}\|_2 w_{n-1}. \end{aligned}$$

Hence, for $n \geq 0$,

(7.41)
$$w_n = -s_{n+1}/\|s_n\|_2.$$

Therefore, one need only store the steps $\{s_n\}$ and their norms to construct the sequence $\{w_n\}$. In fact, we can write (7.38) as

(7.42)
$$B_n^{-1} = \prod_{j=0}^{n-1}\left(I + \frac{s_{j+1}s_j^T}{\|s_j\|_2^2}\right).$$

We cannot use (7.42) and (7.39) directly to compute s_{n+1} because s_{n+1} appears on both sides of the equation

$$
s_{n+1} = -\left(I + \frac{s_{n+1}s_n^T}{\|s_n\|_2^2}\right)\prod_{j=0}^{n-1}\left(I + \frac{s_{j+1}s_j^T}{\|s_j\|_2^2}\right)F(x_{n+1})
$$

(7.43)

$$
= -\left(I + \frac{s_{n+1}s_n^T}{\|s_n\|_2^2}\right)B_n^{-1}F(x_{n+1}).
$$

Instead, we solve (7.43) for s_{n+1} to obtain

(7.44)
$$
s_{n+1} = -\frac{B_n^{-1}F(x_{n+1})}{1 + s_n^T B_n^{-1}F(x_{n+1})/\|s_n\|_2^2}.
$$

By Proposition 7.3.1, the denominator in (7.44)

$$
1 + s_n^T B_n^{-1}F(x_{n+1})/\|s_n\|_2^2 = 1 + v_n^T B_n^{-1}u_n
$$

is nonzero unless B_{n+1} is singular.

The storage requirement, of $n + O(1)$ vectors for the nth iterate, is of the same order as GMRES, making Broyden's method competitive in storage with GMRES as a linear solver [67]. Our implementation is similar to GMRES(m), restarting with B_0 when storage is exhausted.

We can use (7.42) and (7.44) to describe an algorithm. The termination criteria are the same as for Algorithm nsol in § 5.6. We use a restarting approach in the algorithm, adding to the input list a limit $nmax$ on the number of Broyden iterations taken before a restart and a limit $maxit$ on the number of nonlinear iterations. Note that $B_0 = I$ is implicit in the algorithm. Our looping convention is that the loop in step 2(e)ii is skipped if $n = 0$; this is consistent with setting the empty matrix product to the identity. Our notation and algorithmic framework are based on [67].

ALGORITHM 7.3.1. $brsol(x, F, \tau, maxit, nmax)$
1. $r_0 = \|F(x)\|_2$, $n = -1$,
   $s_0 = -F(x)$, $itc = 0$.

2. Do while $itc < maxit$

    (a) $n = n + 1$; $itc = itc + 1$

    (b) $x = x + s_n$

    (c) Evaluate $F(x)$

    (d) If $\|F(x)\|_2 \le \tau_r r_0 + \tau_a$ exit.

    (e) if $n < nmax$ then

        i. $z = -F(x)$

        ii. for $j = 0, n - 1$
            $z = z + s_{j+1}s_j^T z/\|s_j\|_2^2$

> iii. $s_{n+1} = z/(1 - s_n^T z / \|s_n\|_2^2)$

(f) if $n = nmax$ then
> $n = -1; s = -F(x);$

If $n < nmax$, then the nth iteration of brsol, which computes x_n and s_{n+1}, requires $O(nN)$ floating-point operations and storage of $n + 3$ vectors, the steps $\{s_j\}_{j=0}^n$, x, z, and $F(x)$, where z and $F(x)$ may occupy the same storage location. This requirement of one more vector than the algorithm in [67] needed is a result of the nonlinearity. If $n = nmax$, the iteration will restart with x_{nmax}, not compute s_{nmax+1}, and therefore need storage for $nmax + 2$ vectors. Our implementation of brsol in the collection of MATLAB codes stores z and F separately in the interest of clarity and therefore requires $nmax + 3$ vectors of storage.

The Sherman–Morrison approach in Algorithm brsol is more efficient, in terms of both time and storage, than the dense matrix approach proposed in [85]. However the dense matrix approach makes it possible to detect ill conditioning in the approximate Jacobians. If the data is sufficiently good, bounded deterioration implies that the Broyden matrices will be well conditioned and superlinear convergence implies that only a few iterates will be needed. In view of all this, we feel that the approach of Algorithm brsol is the best approach, especially for large problems.

A MATLAB implementation of brsol is provided in the collection of MATLAB codes.

7.4. Examples for Broyden's method

In all the examples in this section we use the l^2 norm multiplied by $1/\sqrt{N}$. We use the MATLAB code brsol.

7.4.1. Linear problems.
The theory in this chapter indicates that Broyden's method can be viewed as an alternative to GMRES as an iterative method for nonsymmetric linear equations. This point has been explored in some detail in [67], where more elaborate numerical experiments that those given here are presented, and in several papers on infinite-dimensional problems as well [91], [104], [120]. As an example we consider the linear PDE (3.33) from § 3.7. We use the same mesh, right-hand side, coefficients, and solution as the example in § 3.7.

We compare Broyden's method without restarts (solid line) to Broyden's method with restarts every three iterates (dashed line). For linear problems, we allow for increases in the nonlinear residual. We set $\tau_a = \tau_r = h^2$ where $h = 1/32$. In Fig. 7.1 we present results for (3.33) preconditioned with the Poisson solver fish2d.

Broyden's method terminated after 9 iterations at a cost of roughly 1.5 million floating point operations, a slightly higher cost than the GMRES solution. The restarted iteration required 24 iterations and 3.5 million floating point operations. One can also see highly irregular performance from the restarted method.

In the linear case [67] one can store only the residuals and recover the terminal iterate upon exit. Storage of the residual, the vector z, and the sequence of steps saves a vector over the nonlinear case.

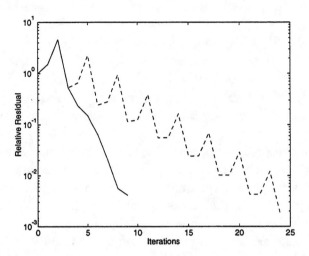

FIG. 7.1. *Broyden's method for the linear PDE.*

7.4.2. Nonlinear problems.

We consider the H-equation from Chapters 5 and 6 and a nonlinear elliptic partial differential equation. Broyden's method and Newton-GMRES both require more storage as iterations progress, Newton-GMRES as the inner iteration progresses and Broyden's method as the outer iteration progresses. The cost in function evaluations for Broyden's method is one for each nonlinear iteration and for Newton-GMRES one for each outer iteration and one for each inner iteration. Hence, if function evaluations are very expensive, the cost of a solution by Broyden's method could be much less than that of one by Newton-GMRES. If storage is at a premium, however, and the number of nonlinear iterations is large, Broyden's method could be at a disadvantage as it might need to be restarted many times. The examples in this section, as well as those in § 8.4 illustrate the differences in the two methods. Our general recommendation is to try both.

Chandrasekhar H-equation. As before we solve the H-equation on a 100-point mesh with $c = .9$ (Fig. 7.2) and $c = .999$ (Fig. 7.3). We compare Broyden's method without restarts (solid line) to Broyden's method with restarts every three iterates (dashed line). For $c = .9$ both methods required six iterations for convergence. The implementation without restarts required 153 thousand floating-point operations and the restarted method 150, a substantial improvement over the Newton-GMRES implementation.

For $c = .9999$, the restarted iteration has much more trouble. The un-restarted iteration converges in 10 iterations at a cost of roughly 249 thousand floating-point operations. The restarted iteration requires 18 iterations and 407 floating-point operations. Even so, both implementations performed better than Newton-GMRES.

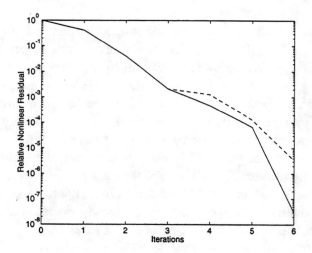

FIG. 7.2. *Broyden's method for the* H-*equation,* $c = .9$.

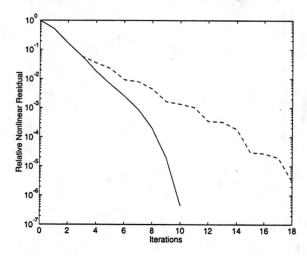

FIG. 7.3. *Broyden's method for the* H-*equation,* $c = .9999$.

One should take some care in drawing general conclusions. While Broyden's method on this example performed better than Newton-GMRES, the storage in Newton-GMRES is used in the inner iteration, while that in Broyden in the nonlinear iteration. If many nonlinear iterations are needed, it may be the case that Broyden's method may need restarts while the GMRES iteration for the linear problem for the Newton step may not.

Convection-diffusion equation.   We consider the nonlinear convection-diffusion equation from § 6.4.2. We use the same mesh width $h = 1/32$ on a uniform square grid. We set $C = 20$, $u_0 = 0$, and $\tau_r = \tau_a = h^2$. We preconditioned with the Poisson solver fish2d.

We allowed for an increase in the residual, which happened on the second iterate. This is not supported by the local convergence theory for nonlinear problems; however it can be quite successful in practice. One can do this in brsol by setting the third entry in the parameter list to 1, as one would if the problem were truly linear. In § 8.4.3 we will return to this example.

In Fig. 7.4 we compare Broyden's method without restarts (solid line) to Broyden's method with restarts every 8 iterates (dashed line). The unrestarted Broyden iteration converges in 12 iterations at a cost of roughly 2.2 million floating-point operations. This is a modest improvement over the Newton-GMRES cost (reported in § 6.4.2) of 2.6 million floating-point operations. However, the residual norms do not monotonically decrease in the Broyden iteration (we fix this in § 8.3.2) and more storage was used.

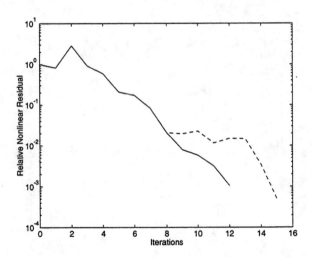

FIG. 7.4.   *Broyden's method for nonlinear PDE.*

At most five inner GMRES iterations were required. Therefore the GMRES inner iteration for the Newton-GMRES approach needed storage for at most 10 vectors (the Krylov basis, s, x_c, $F(x_c)$, $F(x_c + hv)$) to accommodate the Newton-GMRES iteration. The Broyden iteration when restarted every n iterates, requires storage of $n+2$ vectors $\{s_j\}_{j=0}^{n-1}$, x, and z (when stored in the same place as $F(x)$). Hence restarting the Broyden iteration every 8 iterations most closely corresponds to the Newton-GMRES requirements. This approach took 15 iterations to terminate at a cost of 2.4 million floating-point operations, but the convergence was extremely irregular after the restart.

7.5. Exercises on Broyden's method

7.5.1. Prove Proposition 7.2.1.

7.5.2. Prove Proposition 7.3.1.

7.5.3. Extend Proposition 7.3.1 by showing that if B is a nonsingular $N \times N$ matrix, U is $N \times M$, V is $M \times N$, then $A + UV$ is nonsingular if and only if $I + VA^{-1}U$ is a nonsingular $M \times M$ matrix. If this is the case, then

$$(A + UV)^{-1} = A^{-1} - A^{-1}U(I + VA^{-1}U)^{-1}VA^{-1}.$$

This is the *Sherman–Morrison–Woodbury formula* [68], [199]. See [102] for further generalizations.

7.5.4. Duplicate the results reported in § 7.4. Vary the parameters in the equations and the number of vectors stored by Broyden's method and report on their effects on performance. What happens in the differential equation examples if preconditioning is omitted?

7.5.5. Solve the H-equation with Broyden's method and $c = 1$. Set $\tau_a = \tau_r = 10^{-8}$. What q-factor for *linear* convergence do you see? Can you account for this by applying the secant method to the equation $x^2 = 0$? See [53] for a complete explanation.

7.5.6. Try to solve the nonlinear convection-diffusion equation in § 6.4 with Broyden's method using various values for the parameter C. How does the performance of Broyden's method differ from Newton-GMRES?

7.5.7. Modify nsol to use Broyden's method instead of the chord method after the initial Jacobian evaluation. How does this compare with nsol on the examples in Chapter 5.

7.5.8. Compare the performance of Broyden's method and Newton-GMRES on the *Modified Bratu Problem*

$$-\nabla^2 u + du_x + e^u = 0$$

on $(0, 1) \times (0, 1)$ with homogeneous Dirichlet boundary conditions. Precondition with the Poisson solver fish2d. Experiment with various mesh widths, initial iterates, and values for the convection coefficient d.

7.5.9. The *bad Broyden* method [26], so named because of its inferior performance in practice [63], updates an approximation to the inverse of the Jacobian at the root so that $B^{-1} \approx F'(x^*)^{-1}$ satisfies the *inverse secant equation*

(7.45) $$B_+^{-1}y = s$$

with the rank-one update

$$B_+^{-1} = B_c^{-1} + \frac{(s - B_c^{-1}y)y^T}{\|y\|_2^2}.$$

Show that the bad Broyden method is locally superlinearly convergent if the standard assumptions hold. Try to implement the bad Broyden method in a storage-efficient manner using the ideas from [67] and § 7.3. Does anything go wrong? See [67] for more about this.

7.5.10. The *Schubert* or *sparse Broyden* algorithm [172], [27] is a quasi-Newton update for sparse matrices that enforces both the secant equation and the sparsity pattern. For problems with tridiagonal Jacobian, for example, the update is

$$(B_+)_{ij} = (B_c)_{ij} + \frac{(y - B_c s)_i s_j}{\sum_{k=i-1}^{i+1} s_k^2}$$

for $1 \le i, j \le N$ and $|i - j| \le 1$. Note that only the subdiagonal, superdiagonal, and main diagonal are updated. Read about the algorithm in [172] and prove local superlinear convergence under the standard assumptions and the additional assumption that the sparsity pattern of B_0 is the same as that of $F'(x^*)$. How would the Schubert algorithm be affected by preconditioning? Compare your analysis with those in [158], [125], and [189].

7.5.11. Implement the bad Broyden method, apply it to the examples in § 7.4, and compare the performance to that of Broyden's method. Discuss the differences in implementation from Broyden's method.

7.5.12. Implement the Schubert algorithm on an unpreconditioned discretized partial differential equation (such as the Bratu problem from Exercise 7.5.8) and compare it to Newton-GMRES and Broyden's method. Does the relative performance of the methods change as h is decreased? This is interesting even in one space dimension where all matrices are tridiagonal. References [101], [93], [92], and [112], are relevant to this exercise.

7.5.13. Use your result from Exercise 5.7.14 in Chapter 5 to numerically estimate the q-order of Broyden's method for some of the examples in this section (both linear and nonlinear). What can you conclude from your observations?

Chapter 8

Global Convergence

By a *globally convergent algorithm* we mean an algorithm with the property that for any initial iterate the iteration either converges to a root of F or fails to do so in one of a small number of ways. Of the many such algorithms we will focus on the class of *line search* methods and the Armijo rule [3], [88], implemented inexactly [24], [25], [70].

Other methods, such as trust region [153], [154], [181], [129], [63], [24], [25], [173], [70], and continuation/homotopy methods [1], [2], [109], [159], [198], can be used to accomplish the same objective. We select the line search paradigm because of its simplicity, and because it is trivial to add a line search to an existing locally convergent implementation of Newton's method. The implementation and analysis of the line search method we develop in this chapter do not depend on whether iterative or direct methods are used to compute the Newton step or upon how or if the Jacobian is computed and stored. We begin with single equations, where the algorithms can be motivated and intuition developed with a very simple example.

8.1. Single equations

If we apply Newton's method to find the root $x^* = 0$ of the function $f(x) = \arctan(x)$ with initial iterate $x_0 = 10$ we find that the initial iterate is too far from the root for the local convergence theory in Chapter 5 to be applicable. The reason for this is that $f'(x) = (1 + x^2)^{-1}$ is small for large x; $f(x) = \arctan(x) \approx \pm\pi/2$, and hence the magnitude of the Newton step is much larger than that of the iterate itself. We can see this effect in the sequence of iterates:

$$10, -138, 2.9 \times 10^4, -1.5 \times 10^9, 9.9 \times 10^{17},$$

a failure of Newton's method that results from an inaccurate initial iterate.

If we look more closely, we see that the Newton step $s = -f(x_c)/f'(x_c)$ is pointing toward the root in the sense that $sx_c < 0$, but the length of s is too large. This observation motivates the simple fix of reducing the size of the step until the size of the nonlinear residual is decreased. A prototype algorithm is given below.

ALGORITHM 8.1.1. lines1(x, f, τ)

1. $r_0 = |f(x)|$

2. Do while $|f(x)| > \tau_r r_0 + \tau_a$

   (a) If $f'(x) = 0$ terminate with failure.

   (b) $s = -f(x)/f'(x)$ (**search direction**)

   (c) $x_t = x + s$ (**trial point**)

   (d) If $|f(x_t)| < |f(x)|$ then $x = x_t$ (**accept the step**)
   else
   $s = s/2$ goto 2c (**reject the step**)

When we apply Algorithm lines1 to our simple example, we obtain the sequence

$$10, -8.5, 4.9, -3.8, 1.4, -1.3, 1.2, -.99, .56, -0.1, 9 \times 10^{-4}, -6 \times 10^{-10}.$$

The quadratic convergence behavior of Newton's method is only apparent very late in the iteration. Iterates 1, 2, 3, and 4 required 3, 3, 2, and 2 steplength reductions. After the fourth iterate, decrease in the size of the nonlinear residual was obtained with a full Newton step. This is typical of the performance of the algorithms we discuss in this chapter.

We plot the progress of the iteration in Fig. 8.1. We plot

$$|\arctan(x)/\arctan(x_0)|$$

as a function of the number of function evaluations with the solid line. The outer iterations are identified with circles as in § 6.4. One can see that most of the work is in the initial phase of the iteration. In the terminal phase, where the local theory is valid, the minimum number of two function evaluations per iterate is required. However, even in this terminal phase rapid convergence takes place only in the final three outer iterations.

Note the differences between Algorithm lines1 and Algorithm nsol. One does not set $x_+ = x + s$ without testing the step to see if the absolute value of the nonlinear residual has been decreased. We call this method a *line search* because we search along the line segment

$$(x_c, x_c - f(x_c)/f'(x_c))$$

to find a decrease in $|f(x_c)|$. As we move from the right endpoint to the left, some authors [63] refer to this as *backtracking*.

Algorithm lines1 almost always works well. In theory, however, there is the possibility of oscillation about a solution without convergence [63]. To remedy this and to make analysis possible we replace the test for *simple decrease* with one for *sufficient decrease*. The algorithm is to compute a *search direction* d which for us will be the *Newton direction*

$$d = -f(x_c)/f'(x_c)$$

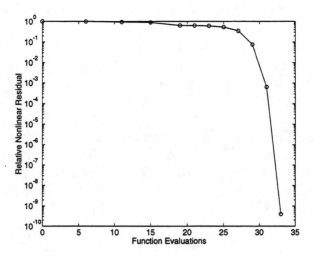

FIG. 8.1. *Newton–Armijo iteration for inverse tangent.*

and then test steps of the form $s = \lambda d$, with $\lambda = 2^{-j}$ for some $j \geq 0$, until $f(x + s)$ satisfies

$$(8.1) \qquad |f(x_c + \lambda d)| < (1 - \alpha\lambda)|f(x_c)|.$$

The condition in (8.1) is called *sufficient decrease* of $|f|$. The parameter $\alpha \in (0,1)$ is a small, but positive, number intended to make (8.1) as easy as possible to satisfy. We follow the recent optimization literature [63] and set $\alpha = 10^{-4}$. Once sufficient decrease has been obtained we *accept the step* $s = \lambda d$. This strategy, from [3] (see also [88]) is called the *Armijo rule*. Note that we must now distinguish between the step and the Newton direction, something we did not have to do in the earlier chapters.

ALGORITHM 8.1.2. nsola1(x, f, τ)

1. $r_0 = |f(x)|$

2. Do while $|f(x)| > \tau_r r_0 + \tau_a$

    (a) If $f'(x) = 0$ terminate with failure.

    (b) $d = -f(x)/f'(x)$ (**search direction**)

    (c) $\lambda = 1$

        i. $x_t = x + \lambda d$ (**trial point**)

        ii. If $|f(x_t)| < (1 - \alpha\lambda)|f(x)|$ then $x = x_t$ (**accept the step**)
        else
        $\lambda = \lambda/2$ goto 2(c)i (**reject the step**)

This is essentially the algorithm implemented in the MATLAB code nsola. We will analyze Algorithm nsola in § 8.2. We remark here that there is nothing critical about the reduction factor of 2 in the line search. A factor of

10 could well be better in situations in which small values of λ are needed for several consecutive steps (such as our example with the arctan function). In that event a reduction of a factor of 8 would require three passes through the loop in nsola1 but only a single reduction by a factor of 10. This could be quite important if the search direction were determined by an inexact Newton method or an approximation to Newton's method such as the chord method or Broyden's method.

On the other hand, reducing λ by too much can be costly as well. Taking full Newton steps ensures fast local convergence. Taking as large a fraction as possible helps move the iteration into the terminal phase in which full steps may be taken and fast convergence expected. We will return to the issue of reduction in λ in § 8.3.

8.2. Analysis of the Armijo rule

In this section we extend the one-dimensional algorithm in two ways before proceeding with the analysis. First, we accept any direction that satisfies the inexact Newton condition (6.1). We write this for the Armijo sequence as

$$(8.2) \qquad \|F'(x_n)d_n + F(x_n)\| \leq \eta_n\|F(x_n)\|.$$

Our second extension is to allow more flexibility in the reduction of λ. We allow for any choice that produces a reduction that satisfies

$$\sigma_0\lambda_{old} \leq \lambda_{new} \leq \sigma_1\lambda_{old},$$

where $0 < \sigma_0 < \sigma_1 < 1$. One such method, developed in § 8.3, is to minimize an interpolating polynomial based on the previous trial steps. The danger is that the minimum may be too near zero to be of much use and, in fact, the iteration may stagnate as a result. The parameter σ_0 safeguards against that. *Safeguarding* is an important aspect of many globally convergent methods and has been used for many years in the context of the polynomial interpolation algorithms we discuss in § 8.3.1 [54], [63], [86], [133], [87]. Typical values of σ_0 and σ_1 are .1 and .5. Our algorithm nsola incorporates these ideas.

While (8.2) is motivated by the Newton-iterative paradigm, the analysis here applies equally to methods that solve the equations for the Newton step directly (so $\eta_n = 0$), approximate the Newton step by a quasi-Newton method, or even use the chord method provided that the resulting direction d_n satisfies (8.2). In Exercise 8.5.9 you are asked to implement the Armijo rule in the context of Algorithm nsol.

ALGORITHM 8.2.1. nsola(x, F, τ, η).

1. $r_0 = \|F(x)\|$

2. Do while $\|F(x)\| > \tau_r r_0 + \tau_a$

    (a) Find d such that $\|F'(x)d + F(x)\| \leq \eta\|F(x)\|$
        If no such d can be found, terminate with failure.

(b) $\lambda = 1$

    i. $x_t = x + \lambda d$

    ii. If $\|F(x_t)\| < (1 - \alpha\lambda)\|F(x)\|$ then $x = x_t$

       else

       Choose $\sigma \in [\sigma_0, \sigma_1]$

       $\lambda = \sigma\lambda$

       goto 2(b)i

Note that step 2a must allow for the possibility that F' is ill-conditioned and that no direction can be found that satisfies (8.2). If, for example, step 2a were implemented with a direct factorization method, ill-conditioning of F' might be detected as part of the factorization.

Let $\{x_n\}$ be the iteration produced by Algorithm nsola with initial iterate x_0. The algorithm can fail in some obvious ways:

1. $F'(x_n)$ is singular for some n. The inner iteration could terminate with a failure.

2. $x_n \to \bar{x}$, a local minimum of $\|F\|$ which is not a root.

3. $\|x_n\| \to \infty$.

Clearly if F has no roots, the algorithm must fail. We will see that if F has no roots then either $F'(x_n)$ has a limit point which is singular or $\{x_n\}$ becomes unbounded.

The convergence theorem is the remarkable statement that if $\{x_n\}$ and $\|F'(x_n)\|^{-1}$ are bounded (thereby eliminating premature failure and local minima of $\|F\|$ that are not roots) and F' is Lipschitz continuous in a neighborhood of *the entire sequence* $\{x_n\}$, then the iteration converges to a root of F at which the standard assumptions hold, full steps are taken in the terminal phase, and the convergence is q-quadratic.

We begin with a formal statement of our assumption that F' is uniformly Lipschitz continuous and bounded away from zero on the sequence $\{x_n\}$.

ASSUMPTION 8.2.1. *There are $r, \gamma, m_f > 0$ such that F is defined, F' is Lipschitz continuous with Lipschitz constant γ, and $\|F'(x)^{-1}\| \leq m_f$ on the set*

$$\Omega(\{x_n\}, r) = \bigcup_{n=0}^{\infty} \{x \mid \|x - x_n\| \leq r\}.$$

Assumption 8.2.1 implies that the line search phase (step 2b in Algorithm **nsola**) will terminate in a number of cycles that is independent of n. We show this, as do [25] and [70], by giving a uniform lower bound on λ_n (assuming that $F(x_0) \neq 0$). Note how the safeguard bound σ_0 enters this result. Note also the very modest conditions on η_n.

LEMMA 8.2.1. *Let $x_0 \in R^N$ and $\alpha \in (0,1)$ be given. Assume that $\{x_n\}$ is given by Algorithm* **nsola** *with*

(8.3) $$\{\eta_n\} \subset (0, \bar{\eta}] \subset (0, 1 - \alpha)$$

and that Assumption 8.2.1 holds. Then either $F(x_0) = 0$ *or*

$$(8.4) \qquad \lambda_n \geq \bar{\lambda} = \sigma_0 \min \left(\frac{r}{m_f \|F(x_0)\|}, \frac{2(1 - \alpha - \bar{\eta})}{m_f^2 \|F(x_0)\|(1 + \bar{\eta})^2 \gamma} \right).$$

Proof. Let $n \geq 0$. By the fundamental theorem of calculus and (8.2) we have, for any $\lambda \in [0, 1]$

$$F(x_n + \lambda d_n) = F(x_n) + \lambda F'(x_n)d_n + \int_0^1 (F'(x_n + t\lambda d_n) - F'(x_n))\lambda d_n \, dt$$

$$= (1 - \lambda)F(x_n) + \lambda \xi_n + \int_0^1 (F'(x_n + t\lambda d_n) - F'(x_n))\lambda d_n \, dt,$$

where, by the bound $\eta_n \leq \bar{\eta}$,

$$\|\xi_n\| \leq \eta_n \|F(x_n)\| \leq \bar{\eta} \|F(x_n)\|.$$

The step acceptance condition in `nsola` implies that $\{\|F(x_n)\|\}$ is a decreasing sequence and therefore

$$\|d_n\| = \|F'(x_n)^{-1}(\xi_n - F(x_n))\| \leq m_f(1 + \bar{\eta})\|F(x_n)\|.$$

Therefore, by Assumption 8.2.1 F' is Lipschitz continuous on the line segment $[x_n, x_n + \lambda d_n]$ whenever

$$\lambda \leq \bar{\lambda}_1 = r/(m_f \|F(x_0)\|).$$

Lipschitz continuity of F' implies that if $\lambda \leq \bar{\lambda}_1$ then

$$\|F(x_n + \lambda d_n)\| \leq (1 - \lambda)\|F(x_n)\| + \lambda \bar{\eta}\|F(x_n)\| + \frac{\gamma}{2}\lambda^2 \|d_n\|^2$$

$$\leq (1 - \lambda)\|F(x_n)\| + \lambda \bar{\eta}\|F(x_n)\| + \frac{m_f^2 \gamma (1 + \bar{\eta})^2 \lambda^2 \|F(x_n)\|^2}{2}$$

$$\leq (1 - \lambda + \bar{\eta}\lambda)\|F(x_n)\| + \lambda\|F(x_n)\| \frac{m_f^2 \gamma \lambda (1 + \bar{\eta})^2 \|F(x_0)\|}{2}$$

$$< (1 - \alpha\lambda)\|F(x_n)\|,$$

which will be implied by

$$\lambda \leq \bar{\lambda}_2 = \min \left(\bar{\lambda}_1, \frac{2(1 - \alpha - \bar{\eta})}{(1 + \bar{\eta})^2 m_f^2 \gamma \|F(x_0)\|} \right).$$

This shows that λ can be no smaller than $\sigma_0 \bar{\lambda}_2$, which completes the proof. $\square$

Now we can show directly that the sequence of Armijo iterates either is unbounded or converges to a solution. Theorem 8.2.1 says even more. The sequence converges to a root at which the standard assumptions hold, so in the terminal phase of the iteration the step lengths are all 1 and the convergence is q-quadratic.

THEOREM 8.2.1. *Let $x_0 \in R^N$ and $\alpha \in (0,1)$ be given. Assume that $\{x_n\}$ is given by Algorithm* **nsola**, *$\{\eta_n\}$ satisfies (8.3), $\{x_n\}$ is bounded, and that Assumption 8.2.1 holds. Then $\{x_n\}$ converges to a root x^* of F at which the standard assumptions hold, full steps are taken for n sufficiently large, and the convergence behavior in the final phase of the iteration is that given by the local theory for inexact Newton methods (Theorem 6.1.2).*

Proof. If $F(x_n) = 0$ for some n, then we are done because Assumption 8.2.1 implies that the standard assumptions hold at $x^* = x_n$. Otherwise Lemma 8.2.1 implies that $F(x_n)$ converges q-linearly to zero with q-factor at most $(1 - \alpha\bar{\lambda})$.

The boundedness of the sequence $\{x_n\}$ implies that a subsequence $\{x_{n_k}\}$ converges, say to x^*. Since F is continuous, $F(x^*) = 0$. Eventually $|x_{n_k} - x^*| < r$, where r is the radius in Assumption 8.2.1 and therefore the standard assumptions hold at x^*.

Since the standard assumptions hold at x^*, there is δ such that if $x \in \mathcal{B}(\delta)$, the Newton iteration with $x_0 = x$ will remain in $\mathcal{B}(\delta)$ and converge q-quadratically to x^*. Hence as soon as $x_{n_k} \in \mathcal{B}(\delta)$, the *entire* sequence, not just the subsequence, will remain in $\mathcal{B}(\delta)$ and converge to x^*. Moreover, Theorem 6.1.2 applies and hence full steps can be taken. $\square$

At this stage we have shown that if the Armijo iteration fails to converge to a root either the continuity properties of F or nonsingularity of F' break down as the iteration progresses (Assumption 8.2.1 is violated) or the iteration becomes unbounded. This is all that one could ask for in terms of robustness of such an algorithm.

Thus far we have used for d the inexact Newton direction. It could well be advantageous to use a chord or Broyden direction, for example. All that one needs to make the theorems and proofs in § 8.2 hold is (8.2), which is easy to verify, in principle, via a forward difference approximation to $F'(x_n)d$.

8.3. Implementation of the Armijo rule

The MATLAB code **nsola** is a modification of **nsolgm** which provides for a choice of several Krylov methods for computing an inexact Newton direction and globalizes Newton's method with the Armijo rule. We use the l^2 norm in that code and in the numerical results reported in § 8.4. If GMRES is used as the linear solver, then the storage requirements are roughly the same as for Algorithm **nsolgm** if one reuses the storage location for the finite difference directional derivative to test for sufficient decrease.

In Exercise 8.5.9 you are asked to do this with **nsol** and, with each iteration, numerically determine if the chord direction satisfies the inexact Newton condition. In this case the storage requirements are roughly the same as for Algorithm **nsol**. The iteration can be very costly if the Jacobian must be evaluated and factored many times during the phase of the iteration in which the approximate solution is far from the root.

We consider two topics in this section. The first, polynomial interpolation as a way to choose λ, is applicable to all methods. The second, how Broyden's method must be implemented, is again based on [67] and is more specialized. Our Broyden–Armijo code does not explicitly check the inexact Newton condition and thereby saves an evaluation of F. Evaluation of $F'(x_c)d_c$ if the full step is not accepted (here d_c is the Broyden direction) would not only verify the inexact Newton condition but provide enough information for a two-point parabolic line search. The reader is asked to pursue this in Exercise 8.5.10.

8.3.1. Polynomial line searches.

In this section we consider an elaboration of the simple line search scheme of reduction of the steplength by a fixed factor. The motivation for this is that some problems respond well to one or two reductions in the steplength by modest amounts (such as $1/2$) and others require many such reductions, but might respond well to a more aggressive steplength reduction (by factors of $1/10$, say). If one can model the decrease in $\|F\|_2$ as the steplength is reduced, one might expect to be able to better estimate the appropriate reduction factor. In practice such methods usually perform better than constant reduction factors.

If we have rejected k steps we have in hand the values

$$\|F(x_n)\|_2, \|F(x_n + \lambda_1 d_n)\|_2, \ldots \|F(x_n + \lambda_{k-1}d_n)\|_2.$$

We can use this iteration history to model the scalar function

$$f(\lambda) = \|F(x_n + \lambda d_n)\|_2^2$$

with a polynomial and use the minimum of that polynomial as the next steplength. We consider two ways of doing this that use second degree polynomials which we compute using previously computed information.

After λ_c has been rejected and a model polynomial computed, we compute the minimum λ_t of that polynomial analytically and set

$$(8.5) \qquad \lambda_+ = \begin{cases} \sigma_0 \lambda_c & \text{if } \lambda_t < \sigma_0 \lambda_c, \\[2mm] \sigma_1 \lambda_c & \text{if } \lambda_t > \sigma_1 \lambda_c, \\[2mm] \lambda_t & \text{otherwise.} \end{cases}$$

Two-point parabolic model. In this approach we use values of f and f' at $\lambda = 0$ and the value of f at the current value of λ to construct a 2nd degree interpolating polynomial for f.

$f(0) = \|F(x_n)\|_2^2$ is known. Clearly $f'(0)$ can be computed as

$$(8.6) \qquad f'(0) = 2(F'(x_n)^T d_n)^T F(x_n) = 2F(x_n)^T (F'(x_n)d_n).$$

Use of (8.6) requires evaluation of $F'(x_n)d_n$, which can be obtained by examination of the final residual from GMRES or by expending an additional function evaluation to compute $F'(x_n)d_n$ with a forward difference. In any case, our approximation to $f'(0)$ should be negative. If it is not, it may be necessary to compute a new search direction. This is a possibility with directions other than inexact Newton directions, such as the Broyden direction.

The polynomial $p(\lambda)$ such that p, p' agree with f, f' at 0 and $p(\lambda_c) = f(\lambda_c)$ is

$$p(\lambda) = f(0) + f'(0)\lambda + c\lambda^2,$$

where

$$c = \frac{f(\lambda_c) - f(0) - f'(0)\lambda_c}{\lambda_c^2}.$$

Our approximation of $f'(0) < 0$, so if $f(\lambda_c) > f(0)$, then $c > 0$ and $p(\lambda)$ has a minimum at

$$\lambda_t = -f'(0)/(2c) > 0.$$

We then compute λ_+ with (8.5).

Three-point parabolic model. An alternative to the two-point model that avoids the need to approximate $f'(0)$ is a three-point model, which uses $f(0)$ and the two most recently rejected steps to create the parabola. The MATLAB code parab3p implements the three-point parabolic line search and is called by the two codes nsola and brsola.

In this approach one evaluates $f(0)$ and $f(1)$ as before. If the full step is rejected, we set $\lambda = \sigma_1$ and try again. After the second step is rejected, we have the values

$$f(0), f(\lambda_c), \text{ and } f(\lambda_-),$$

where λ_c and λ_- are the most recently rejected values of λ. The polynomial that interpolates f at $0, \lambda_c, \lambda_-$ is

$$p(\lambda) = f(0) + \frac{\lambda}{\lambda_c - \lambda_-} \left(\frac{(\lambda - \lambda_-)(f(\lambda_c) - f(0))}{\lambda_c} + \frac{(\lambda_c - \lambda)(f(\lambda_-) - f(0))}{\lambda_-} \right).$$

We must consider two situations. If

$$p''(0) = \frac{2}{\lambda_c \lambda_- (\lambda_c - \lambda_-)} (\lambda_-(f(\lambda_c) - f(0)) - \lambda_c(f(\lambda_-) - f(0)))$$

is positive, then we set λ_t to the minimum of p

$$\lambda_t = -p'(0)/p''(0)$$

and apply the safeguarding step (8.5) to compute λ_+. If $p''(0) \le 0$ one could either set λ_+ to be the minimum of p on the interval $[\sigma_0\lambda, \sigma_1\lambda]$ or reject the parabolic model and simply set λ_+ to $\sigma_0\lambda_c$ or $\sigma_1\lambda_c$. We take the latter approach and set $\lambda_+ = \sigma_1\lambda_c$. In the MATLAB code nsola from the collection, we implement this method with $\sigma_0 = .1$, $\sigma_1 = .5$.

Interpolation with a cubic polynomial has also been suggested [87], [86], [63], [133]. Clearly, the polynomial modeling approach is heuristic. The safeguarding and Theorem 8.2.1 provide insurance that progress will be made in reducing $\|F\|$.

8.3.2. Broyden's method.
In Chapter 7 we implemented Broyden's method at a storage cost of one vector for each iteration. The relation $y - B_c s = F(x_+)$ was not critical to this and we may also incorporate a line search at a cost of a bit of complexity in the implementation. As in § 7.3, our approach is a nonlinear version of the method in [67].

The difference from the development in § 7.3 is that the simple relation between the sequence $\{w_n\}$ and $\{v_n\}$ is changed. If a line search is used, then

$$s_n = -\lambda_n B_n^{-1} F(x_n)$$

and hence

(8.7) $\qquad y_n - B_n s_n = F(x_{n+1}) - (1 - \lambda_n) F(x_n).$

If, using the notation in § 7.3, we set

$$u_n = \frac{y_n - B_n s_n}{\|s_n\|_2}, v_n = \frac{s_n}{\|s_n\|_2}, \text{ and } w_n = (B_n^{-1} u_n)/(1 + v_n^T B_n^{-1} u_n),$$

we can use (8.7) and (7.38) to obtain

(8.8)
$$
\begin{aligned}
w_n &= \lambda_n \left(\frac{-d_{n+1}}{\|s_n\|_2} + (\lambda_n^{-1} - 1) \frac{s_n}{\|s_n\|_2} \right) \\
&= \lambda_n \left(\frac{-s_{n+1}/\lambda_{n+1}}{\|s_n\|_2} + (\lambda_n^{-1} - 1) \frac{s_n}{\|s_n\|_2} \right),
\end{aligned}
$$

where

$$d_{n+1} = -B_{n+1}^{-1} F(x_{n+1})$$

is the search direction used to compute x_{n+1}. Note that (8.8) reduces to (7.41) when $\lambda_n = 1$ and $\lambda_{n+1} = 1$ (so $d_{n+1} = s_{n+1}$).

We can use the first equality in (8.8) and the relation

$$d_{n+1} = - \left(I - \frac{w_n s_n^T}{\|s_n\|_2} \right) B_n^{-1} F(x_{n+1})$$

to solve for d_{n+1} and obtain an analog of (7.44)

(8.9) $\qquad d_{n+1} = - \dfrac{\|s_n\|_2^2 B_n^{-1} F(x_{n+1}) - (1 - \lambda_n) s_n^T B_n^{-1} F(x_{n+1}) s_n}{\|s_n\|_2^2 + \lambda_n s_n^T B_n^{-1} F(x_{n+1})}$

We then use the second equality in (8.8) to construct $B_c^{-1} F(x_+)$ as we did in Algorithm brsol. Verification of (8.8) and (8.9) is left to Exercise 8.5.5.

Algorthm `brsola` uses these ideas. We do not include details of the line search or provide for restarts or a limited memory implementation. The MATLAB implementation, provided in the collection of MATLAB codes, uses a three-point parabolic line search and performs a restart when storage is exhausted like Algorithm `brsol` does.

It is possible that the Broyden direction fails to satisfy (8.2). In fact, there is no guarantee that the Broyden direction is a direction of decrease for $\|F\|_2$. The MATLAB code returns with an error message if more than a fixed number of reductions in the step length are needed. Another approach would be to compute $F'(x_c)d$ numerically if a full step is rejected and then test (8.2) before beginning the line search. As a side effect, an approximation of $f'(0)$ can be obtained at no additional cost and a parabolic line search based on $f(0)$, $f'(0)$, and $f(1)$ can be used. In Exercise 8.5.10 the reader is asked to fully develop and implement these ideas.

ALGORITHM 8.3.1. $\mathtt{brsola}(x, F, \tau, maxit, nmax)$

1. Initialize:

   $F_c = F(x)$ $r_0 = \|F(x)\|_2$, $n = -1$,
   $d = -F(x)$, compute λ_0, $s_0 = \lambda_0 d$

2. Do while $n \leq maxit$

   (a) $n = n + 1$

   (b) $x = x + s_n$

   (c) Evaluate $F(x)$

   (d) If $\|F(x)\|_2 \leq \tau_r r_0 + \tau_a$ exit.

   (e)   i. $z = -F(x)$

         ii. for $j = 0, n - 1$
             $a = -\lambda_j / \lambda_{j+1}$, $b = 1 - \lambda_j$
             $z = z + (a s_{j+1} + b s_j) s_j^T z / \|s_j\|_2^2$

   (f) $d = (\|s_n\|_2^2 z + (1 - \lambda_n) s_n^T z s_n)/(\|s_n\|_2^2 - \lambda_n s_n^T z)$

   (g) Compute λ_{n+1} with a line search.

   (h) Set $s_{n+1} = \lambda_{n+1} d$, store $\|s_{n+1}\|_2$ and λ_{n+1}.

Since d_{n+1} and s_{n+1} can occupy the same location, the storage requirements of Algorithm `brsola` would appear to be essentially the same as those of `brsol`. However, computation of the step length requires storage of both the current point x and the trial point $x + \lambda s$ before a step can be accepted. So, the storage of the globalized methods exceeds that of the local method by one vector. The MATLAB implementation of `brsola` in the collection illustrates this point.

8.4. Examples for Newton–Armijo

The MATLAB code `nsola` is a modification of `nsolgm` that incorporates the three-point parabolic line search from § 8.3.1 and also changes η using (6.20) once the iteration is near the root. We compare this strategy, using GMRES as the linear solver, with the constant η method as we did in § 6.4 with $\gamma = .9$.

As before, in all the figures we plot $\|F(x_n)\|_2/\|F(x_0)\|_2$ against the number of function evaluations required by all line searches and inner and outer iterations to compute x_n. Counts of function evaluations corresponding to outer iterations are indicated by circles. We base our absolute error criterion on the norm $\| \cdot \|_2/\sqrt{N}$ as we did in § 6.4.2 and 7.4.

In § 8.4.3 we compare the Broyden–Armijo method with Newton-GMRES for those problems for which both methods were successful. We caution the reader now, and will repeat the caution later, that if the initial iterate is far from the solution, an inexact Newton method such as Newton-GMRES can succeed in many case in which a quasi-Newton method can fail because the quasi-Newton direction may not be an inexact Newton direction. However, when both methods succeed, the quasi-Newton method, which requires a single function evaluation for each outer iteration when full steps can be taken, may well be most efficient.

8.4.1. Inverse tangent function.

Since we have easy access to analytic derivatives in this example, we can use the two-point parabolic line search. We compare the two-point parabolic line search with the constant reduction search ($\sigma_0 = \sigma_1 = .5$) for the arctan function. In Fig. 8.2 we plot the iteration history corresponding to the parabolic line search with the solid line and that for the constant reduction with the dashed line. We use $x_0 = 10$ as the initial iterate with $\tau_r = \tau_a = 10^{-8}$. The parabolic line search required 7 outer iterates and 21 function evaluations in contrast with the constant reduction searches 11 outer iterates and 33 function evaluations. In the first outer iteration, both line searches take three steplength reductions. However, the parabolic line search takes only one reduction in the next three iterations and none thereafter. The constant reduction line search took three reductions in the first two outer iterations and two each in following two.

8.4.2. Convection-diffusion equation.

We consider a much more difficult problem. We take (6.21) from § 6.4.2,

$$-\nabla^2 u + Cu(u_x + u_y) = f$$

with the same right hand side f, initial iterate $u_0 = 0$, 31×31 mesh and homogeneous Dirichlet boundary conditions on the unit square $(0,1) \times (0,1)$ that we considered before. Here, however, we set $C = 100$. This makes a globalization strategy critical (Try it without one!). We set $\tau_r = \tau_a = h^2/10$. This is a tighter tolerance that in § 6.4.2 and, because of the large value of C and resulting ill-conditioning, is needed to obtain an accurate solution.

In Fig. 8.3 we show the progress of the iteration for the unpreconditioned equation. For this problem we plot the progress of the iteration using $\eta = .25$ with the solid line and using the sequence given by (6.20) with the dashed

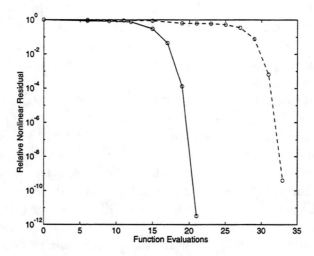

FIG. 8.2. *Newton–Armijo for the* arctan *function.*

line. We set the parameters in (6.20) to $\gamma = .9$ and $\eta_{max} = .25$. This problem is very poorly conditioned and much tighter control on the inner iteration is needed than for the other problems. The two approaches to selection of the forcing term performed very similarly. The iterations required 75.3 (constant η) and 75.4 (varying η) million floating point operations, 759 (constant η) and 744 (varying η) function evaluations, and 25 (constant η) and 22 (varying η) outer iterations.

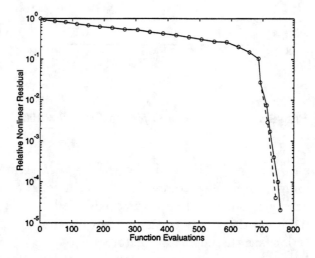

FIG. 8.3. *Newton–Armijo for the PDE, $C = 100$.*

We consider the preconditioned problem in Fig. 8.4. In the computation we preconditioned (6.21) with the fast Poisson solver fish2d. In Fig. 8.4 we show the progress of the iteration for the preconditioned equation. For this problem we plot the progress of the iteration using $\eta = .25$ with the solid line and using the sequence given by (6.20) with the dashed line. We set the parameters in (6.20) to $\gamma = .9$ and $\eta_{max} = .99$.

The constant η iteration terminated after 79 function evaluations, 9 outer iterations, and roughly 13.5 million floating-point operations. The line search in this case reduced the step length three times on the first iteration and twice on the second and third.

The iteration in which $\{\eta_n\}$ is given by (6.20) terminated after 70 function evaluations, 9 outer iterations, and 12.3 million floating-point operations. The line search in the non-constant η case reduced the step length three times on the first iteration and once on the second. The most efficient iteration, with the forcing term given by (6.20), required at most 16 inner iterations while the constant η approach needed at most 10.

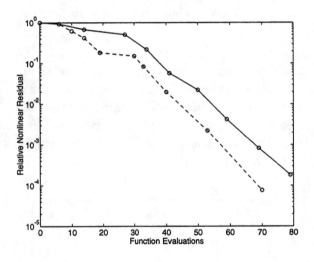

FIG. 8.4. *Newton–Armijo for the PDE, $C = 100$.*

8.4.3. Broyden–Armijo.

We found that the Broyden–Armijo line search failed on all the unpreconditioned convection diffusion equation examples. In these cases the Jacobian is not well approximated by a low rank perturbation of the identity [112] and the ability of the inexact Newton iteration to find a good search direction was the important factor.

We begin by revisiting the example in § 6.4.2 with $C = 20$. As reported in § 6.4.2, the Newton-GMRES iteration always took full steps, terminating in 4 outer iterations, 16 function evaluations, and 2.6 million floating-point operations. When compared to the Broyden's method results in § 6.4.2, when increases in the residual were allowed, the Broyden–Armijo costs reflect an improvement in efficiency. The Broyden iteration in § 6.4.2 took 12 iterations and 2.8 million floating-point operations while the Broyden–Armijo took 9 iterations and required 2 million floating-point operations. We set $\tau_a = \tau_r = h^2$, $\eta_{max} = .5$, and $\gamma = .9$ as we did in § 6.4.2.

In our implementation of brsola the three-point parabolic line search was used. The steplength was reduced on the second iteration (twice) and the third (once). Figure 8.5 compares the Broyden–Armijo iteration (solid line) to the GMRES iteration (dashed line).

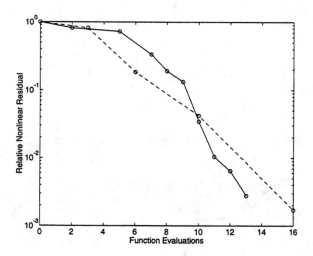

FIG. 8.5. *Newton-GMRES and Broyden–Armijo for the PDE, $C = 20$.*

For the more difficult problem with $C = 100$, the performance of the two methods is more similar. In Fig. 8.6 we compare our implementation of Newton–GMRES–Armijo (dashed line) to Broyden–Armijo (solid line). We set $\tau_a = \tau_r = h^2/10$, $\eta_{max} = .99$, and $\gamma = .9$. The Broyden–Armijo iteration required 34 nonlinear iterations, roughly 16 million floating-point operations, and 85 function evaluations. In terms of storage, the Broyden–Armijo iteration required storage for 37 vectors while the Newton-GMRES iteration needed at most 16 inner iterations, needing to store 21 vectors, and therefore was much more efficient in terms of storage.

Restarting the Broyden iteration every 19 iterations would equalize the storage requirements with Newton–GMRES–Armijo. Broyden's method suffers under these conditions as one can see from the comparison of Broyden–Armijo (solid line) and Newton–GMRES–Armijo (dashed line) in Fig. 8.7. The restarted Broyden–Armijo iteration took 19.5 million floating-point operations, 42 outer iterates, and 123 function evaluations.

From these examples we see that Broyden–Armijo can be very efficient provided only a small number of iterations are needed. The storage requirements increase as the nonlinear iteration progresses and this fact puts Broyden's

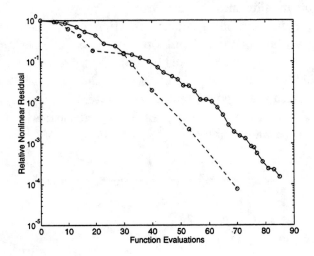

FIG. 8.6.   *Newton–GMRES and Broyden–Armijo for the PDE, $C = 100$.*

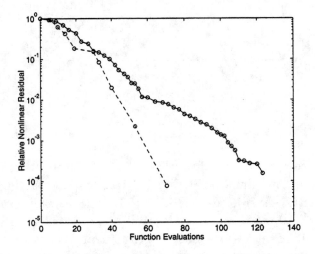

FIG. 8.7.   *Newton–GMRES and restarted Broyden–Armijo for the PDE, $C = 100$.*

method at a disadvantage to Newton-GMRES when the number of nonlinear
iterations is large.

8.5. Exercises on global convergence

8.5.1. How does the method proposed in [10] differ from the one implemented in nsola? What could be advantages and disadvantages of that approach?

8.5.2. In what ways are the results in [25] and [70] more general than those in this section? What ideas do these papers share with the analysis in this section and with [3] and [88]?

8.5.3. Implement the Newton–Armijo method for single equations. Apply your code to the following functions with $x_0 = 10$. Explain your results.

   1. $f(x) = \arctan(x)$ (i.e., Duplicate the results in § 8.1.)

   2. $f(x) = \arctan(x^2)$

   3. $f(x) = .9 + \arctan(x)$

   4. $f(x) = x(1 + \sin(1/x))$

   5. $f(x) = e^x$

   6. $f(x) = 2 + \sin(x)$

   7. $f(x) = 1 + x^2$

8.5.4. A numerical analyst buys a German sports car for $50,000. He puts $10,000 down and takes a 7 year installment loan to pay the balance. If the monthly payments are $713.40, what is the interest rate? Assume monthly compounding.

8.5.5. Verify (8.8) and (8.9).

8.5.6. Show that if F' is Lipschitz continuous and the iteration $\{x_n\}$ produced by Algorithm nsola converges to x^* with $F(x^*) \neq 0$, then $F'(x^*)$ is singular.

8.5.7. Use nsola to duplicate the results in § 8.4.2. Vary the convection coefficient in the convection-diffusion equation and the mesh size and report the results.

8.5.8. Experiment with other linear solvers such as GMRES(m), Bi-CGSTAB, and TFQMR. This is interesting in the locally convergent case as well. You might use the MATLAB code nsola to do this.

8.5.9. Modify nsol, the hybrid Newton algorithm from Chapter 5, to use the Armijo rule. Try to do it in such a way that the chord direction is used whenever possible.

8.5.10. Modify brsola to test the Broyden direction for the descent property and use a two-point parabolic line search. What could you do if the Broyden direction is not a descent direction? Apply your code to the examples in § 8.4.

8.5.11. Modify nsola to use a cubic polynomial and constant reduction line searches instead of the quadratic polynomial line search. Compare the results with the examples in § 8.4.

8.5.12. Does the secant method for equations in one variable always give a direction that satisfies (8.2) with η_n bounded away from 1? If not, when does it? How would you implement a secant-Armijo method in such a way that the convergence theorem 8.2.1 is applicable?

8.5.13. Under what conditions will the iteration given by nsola converge to a root x^* that is independent of the initial iterate?

Bibliography

[1] E. L. ALLGOWER AND K. GEORG, *Simplicial and continuation methods for approximating fixed points and solutions to systems of equations*, SIAM Rev., 22 (1980), pp. 28–85.

[2] ——, *Numerical Continuation Methods : An Introduction*, Springer-Verlag, New York, 1990.

[3] L. ARMIJO, *Minimization of functions having Lipschitz-continuous first partial derivatives*, Pacific J. Math., 16 (1966), pp. 1–3.

[4] W. E. ARNOLDI, *The principle of minimized iterations in the solution of the matrix eigenvalue problem*, Quart. Appl. Math., 9 (1951), pp. 17–29.

[5] S. F. ASHBY, T. A. MANTEUFFEL, AND J. S. OTTO, *A comparison of adaptive Chebyshev and least squares polynomial preconditioning for Hermetian positive definite linear systems*, SIAM J. Sci. Statist. Comput., 13 (1992), pp. 1–29.

[6] K. E. ATKINSON, *Iterative variants of the Nyström method for the numerical solution of integral equations*, Numer. Math., 22 (1973), pp. 17–31.

[7] ——, *An Introduction to Numerical Analysis*, 2nd. ed., John Wiley, New York, 1989.

[8] O. AXELSSON, *Iterative Solution Methods*, Cambridge University Press, Cambridge, 1994.

[9] S. BANACH, *Sur les opérations dans les ensembles abstraits et leur applications aux équations intégrales*, Fund. Math, 3 (1922), pp. 133–181.

[10] R. E. BANK AND D. J. ROSE, *Global approximate Newton methods*, Numer. Math., 37 (1981), pp. 279–295.

[11] M. S. BARLETT, *An inverse matrix adjustment arising in discriminant analysis*, Ann. Math. Statist., 22 (1951), pp. 107–111.

[12] R. BARRETT, M. BERRY, T. F. CHAN, J. DEMMEL, J. DONATO, J. DONGARRA, V. EIJKHOUT, R. POZO, C. ROMINE, AND H. VAN DER VORST, *Templates for the Solution of Linear Systems: Building Blocks for Iterative Methods*, Society for Industrial and Applied Mathematics, Philadelphia, PA, 1993.

[13] N. BIĆANIĆ AND K. H. JOHNSON, *Who was '-Raphson'?*, Internat. J. Numer. Methods. Engrg., 14 (1979), pp. 148–152.

[14] P. B. BOSMA AND W. A. DEROOIJ, *Efficient methods to calculate Chandrasekhar's H-functions*, Astron. Astrophys. Lib., 126 (1983), p. 283.

[15] H. BRAKHAGE, *Über die numerische Behandlung von Integralgleichungen nach der Quadraturformelmethode*, Numer. Math., 2 (1960), pp. 183–196.

[16] K. E. BRENAN, S. L. CAMPBELL, AND L. R. PETZOLD, *Numerical Solution*

of Initial Value Problems in Differential-Algebraic Equations, no. 14 in Classics in Applied Mathematics, SIAM, Philadelphia, 1996.

[17] R. BRENT, *Algorithms for Minimization Without Deriviatives*, Prentice-Hall, Englewood Cliffs, NJ, 1973.

[18] ——, *Some efficient algorithms for solving systems of nonlinear equations*, SIAM J. Numer. Anal., 10 (1973), pp. 327–344.

[19] W. BRIGGS, *A Multigrid Tutorial*, Society for Industrial and Applied Mathematics, Philadelphia, PA, 1987.

[20] P. N. BROWN, *A local convergence theory for combined inexact–Newton/ finite-difference projection methods*, SIAM J. Numer. Anal., 24 (1987), pp. 407–434.

[21] P. N. BROWN, G. D. BYRNE, AND A. C. HINDMARSH, *VODE: A variable coefficient ode solver*, SIAM J. Sci. Statist. Comput., 10 (1989), pp. 1038–1051.

[22] P. N. BROWN AND A. C. HINDMARSH, *Reduced storage matrix methods in stiff ODE systems*, J. Appl. Math. Comput., 31 (1989), pp. 40–91.

[23] P. N. BROWN, A. C. HINDMARSH, AND L. R. PETZOLD, *Using Krylov methods in the solution of large-scale differential-algebraic systems*, SIAM J. Sci. Comput., 15 (1994), pp. 1467–1488.

[24] P. N. BROWN AND Y. SAAD, *Hybrid Krylov methods for nonlinear systems of equations*, SIAM J. Sci. Statist. Comput., 11 (1990), pp. 450–481.

[25] ——, *Convergence theory of nonlinear Newton-Krylov algorithms*, SIAM J. Optim., 4 (1994), pp. 297–330.

[26] C. G. BROYDEN, *A class of methods for solving nonlinear simultaneous equations*, Math. Comput., 19 (1965), pp. 577–593.

[27] ——, *The convergence of an algorithm for solving sparse nonlinear systems*, Math. Comput., 25 (1971), pp. 285–294.

[28] C. G. BROYDEN, J. E. DENNIS, AND J. J. MORÉ, *On the local and superlinear convergence of quasi-Newton methods*, J. Inst. Maths. Applic., 12 (1973), pp. 223–246.

[29] W. BURMEISTER, *Zur Konvergenz einiger verfahren der konjugierten Richtungen*, in Proceedings of Internationaler Kongreß über Anwendung der Mathematik in dem Ingenieurwissenschaften, Weimar, 1975.

[30] I. W. BUSBRIDGE, *The Mathematics of Radiative Transfer*, Cambridge Tracts, No. 50, Cambridge Univ. Press, Cambridge, 1960.

[31] R. H. BYRD, J. NOCEDAL, AND R. B. SCHNABEL, *Representation of quasi-Newton matrices and their use in limited memory methods*, Math. Programming, 63 (1994), pp. 129–156.

[32] X.-C. CAI, W. D. GROPP, D. E. KEYES, AND M. D. TIDRIRI, *Newton-Krylov-Schwarz methods in CFD*, in Proceedings of the International Workshop on the Navier-Stokes Equations, R. Rannacher, ed., Notes in Numerical Fluid Mechanics, Braunschweig, 1994, Vieweg Verlag.

[33] S. L. CAMPBELL, I. C. F. IPSEN, C. T. KELLEY, AND C. D. MEYER, *GMRES and the minimal polynomial*, Tech. Report CRSC-TR94-10, North Carolina State University, Center for Research in Scientific Computation, July 1994. BIT, to appear.

[34] S. L. CAMPBELL, I. C. F. IPSEN, C. T. KELLEY, C. D. MEYER, AND Z. Q. XUE, *Convergence estimates for solution of integral equations with GMRES*, Tech. Report CRSC-TR95-13, North Carolina State University, Center for Research in Scientific Computation, March 1995. Journal of Integral Equations and Applications, to appear.

[35] R. CAVANAUGH, *Difference Equations and Iterative Processes*, PhD thesis, University of Maryland, 1970.

[36] F. CHAITIN-CHATELIN, *Is nonnormality a serious difficulty ?*, Tech. Report TR/PA/94/18, CERFACS, December 1994.

[37] T. CHAN, E. GALLOPOULOS, V. SIMONCINI, T. SZETO, AND C. TONG, *A quasi-minimal residual variant of the Bi-CGSTAB algorithm for nonsymmetric systems*, SIAM J. Sci. Comput., 15 (1994), p. 338.

[38] T. CHAN, R. GLOWINSKI, J. PÉRIAUX, AND O. WIDLUND, eds., *Domain Decomposition Methods*,Proceedings of the Second International Symposium on Domain Decomposition Methods, Los Angeles, CA, January 14–16, 1988; Society for Industrial and Applied Mathematics, Philadelphia, PA, 1989.

[39] ——, eds., *Domain Decomposition Methods*, Proceedings of the Third International Symposium on Domain Decomposition Methods, Houston, TX, 1989; Society for Industrial and Applied Mathematics, Philadelphia, PA, 1990.

[40] ——, eds., *Domain Decomposition Methods*, Proceedings of the Fourth International Symposium on Domain Decomposition Methods, Moscow, USSR, 1990; Society for Industrial and Applied Mathematics, Philadelphia, PA, 1991.,

[41] S. CHANDRASEKHAR, *Radiative Transfer*, Dover, New York, 1960.

[42] T. F. COLEMAN AND J. J. MORÉ, *Estimation of sparse Jacobian matrices and graph coloring problems*, SIAM J. Numer. Anal., 20 (1983), pp. 187–209.

[43] T. F. COLEMAN AND C. VANLOAN, *Handbook for Matrix Computations*, Frontiers in Appl. Math., No. 4, Society for Industrial and Applied Mathematics, Philadelphia, PA, 1988.

[44] P. CONCUS, G. H. GOLUB, AND G. MEURANT, *Block preconditioning for the conjugate gradient method*, SIAM J. Sci. Statist. Comput., 6 (1985), pp. 220–252.

[45] P. CONCUS, G. H. GOLUB, AND D. P. O'LEARY, *A generalized conjugate gradient method for the numerical solution of elliptic partial differential equations*, in Sparse Matrix Computations, J. R. Bunch and D. J. Rose, eds., Academic Press, 1976, pp. 309–332.

[46] E. J. CRAIG, *The N-step iteration procedures*, J. Math. Phys., 34 (1955), pp. 64–73.

[47] A. R. CURTIS, M. J. D. POWELL, AND J. K. REID, *On the estimation of sparse Jacobian matrices*, J. Inst. Math. Appl., 13 (1974), pp. 117–119.

[48] J. W. DANIEL, *The conjugate gradient method for linear and nonlinear operator equations*, SIAM J. Numer. Anal., 4 (1967), pp. 10–26.

[49] D. W. DECKER, H. B. KELLER, AND C. T. KELLEY, *Convergence rates for Newton's method at singular points*, SIAM J. Numer. Anal., 20 (1983), pp. 296–314.

[50] D. W. DECKER AND C. T. KELLEY, *Newton's method at singular points I*, SIAM J. Numer. Anal., 17 (1980), pp. 66–70.

[51] ——, *Convergence acceleration for Newton's method at singular points*, SIAM J. Numer. Anal., 19 (1982), pp. 219–229.

[52] ——, *Sublinear convergence of the chord method at singular points*, Numer. Math., 42 (1983), pp. 147–154.

[53] ——, *Broyden's method for a class of problems having singular Jacobian at the root*, SIAM J. Numer. Anal., 22 (1985), pp. 566–574.

[54] T. J. DEKKER, *Finding a zero by means of successive linear interpolation*, in Constructive Aspects of the Fundamental Theorem of Algebra, P. Henrici, ed., 1969, pp. 37–48.

[55] R. DEMBO, S. EISENSTAT, AND T. STEIHAUG, *Inexact Newton methods*, SIAM J. Numer. Anal., 19 (1982), pp. 400–408.

[56] R. DEMBO AND T. STEIHAUG, *Truncated Newton algorithms for large-scale optimization*, Math. Programming, 26 (1983), pp. 190–212.

[57] J. E. DENNIS, *On the Kantorovich hypothesis for Newton's method*, SIAM J. Numer. Anal., 6 (1969), pp. 493–507.

[58] ——, *Toward a unified convergence theory for Newton-like methods*, in Nonlinear Functional Analysis and Applications, L. B. Rall, ed., Academic Press, New York, 1971, pp. 425–472.

[59] J. E. DENNIS, J. M. MARTINEZ, AND X. ZHANG, *Triangular decomposition methods for solving reducible nonlinear systems of equations*, SIAM J. Optim., 4 (1994), pp. 358–382.

[60] J. E. DENNIS AND J. J. MORÉ, *A characterization of superlinear convergence and its application to quasi-Newton methods*, Math. Comput., 28 (1974), pp. 549–560.

[61] ——, *Quasi-Newton methods, methods, motivation and theory*, SIAM Rev., 19 (1977), pp. 46–89.

[62] J. E. DENNIS AND R. B. SCHNABEL, *Least change secant updates for quasi-Newton methods*, SIAM Rev., 21 (1979), pp. 443–459.

[63] ——, *Numerical Methods for Unconstrained Optimization and Nonlinear Equations*, no. 16 in Classics in Applied Mathematics, SIAM, Philadelphia, 1996.

[64] J. E. DENNIS AND H. F. WALKER, *Convergence theorems for least change secant update methods*, SIAM J. Numer. Anal., 18 (1981), pp. 949–987.

[65] ——, *Inaccuracy in quasi-Newton methods: Local improvement theorems*, in Mathematical Programming Study 22: Mathematical programming at Oberwolfach II, North–Holland, Amsterdam, 1984, pp. 70–85.

[66] ——, *Least-change sparse secant updates with inaccurate secant conditions*, SIAM J. Numer. Anal., 22 (1985), pp. 760–778.

[67] P. DEUFLHARD, R. W. FREUND, AND A. WALTER, *Fast secant methods for the iterative solution of large nonsymmetric linear systems*, Impact of Computing in Science and Engineering, 2 (1990), pp. 244–276.

[68] W. J. DUNCAN, *Some devices for the solution of large sets of simultaneous linear equations (with an appendix on the reciprocation of partitioned matrices)*, The London, Edinburgh, and Dublin Philosophical Magazine and Journal of Science, Seventh Series, 35 (1944), pp. 660–670.

[69] S. C. EISENSTAT AND H. F. WALKER, *Choosing the forcing terms in an inexact Newton method*, SIAM J. Sci. Comput., 17 (1996), pp. 16–32.

[70] ——, *Globally convergent inexact Newton methods*, SIAM J. Optim., 4 (1994), pp. 393–422.

[71] H. C. ELMAN, *Iterative Methods for Large, Sparse, Nonsymmetric Systems of Linear Equations*, PhD thesis, Yale University, New Haven, CT, 1982.

[72] H. C. ELMAN, Y. SAAD, AND P. E. SAYLOR, *A hybrid Chebyshev-Krylov subspace algorithm for solving nonsymmetric systems of linear equations*, SIAM J. Sci. Statist. Comput., 7 (1986), pp. 840–855.

[73] M. ENGELMAN, G. STRANG, AND K. J. BATHE, *The application of quasi-Newton methods in fluid mechanics*, Internat. J. Numer. Methods Engrg., 17 (1981), pp. 707–718.

[74] V. FABER AND T. A. MANTEUFFEL, *Necessary and sufficient conditions for the existence of a conjugate gradient method*, SIAM J. Numer. Anal., 21 (1984), pp. 352–362.

[75] D. FENG, P. D. FRANK, AND R. B. SCHNABEL, *Local convergence analysis of tensor methods for nonlinear equations*, Math. Programming, 62 (1993), pp. 427–459.

[76] R. FLETCHER, *Conjugate gradient methods for indefinite systems*, in Numerical Analysis Dundee 1975, G. Watson, ed., Springer-Verlag, Berlin, New York, 1976, pp. 73–89.

[77] R. W. FREUND, *A transpose-free quasi-minimal residual algorithm for non-Hermitian linear systems*, SIAM J. Sci. Comput., 14 (1993), pp. 470–482.

[78] R. W. FREUND, G. H. GOLUB, AND N. M. NACHTIGAL, *Iterative solution of linear systems*, Acta Numerica, 1 (1991), pp. 57–100.

[79] R. W. FREUND, M. H. GUTKNECHT, AND N. M. NACHTIGAL, *An implementation of the look-ahead Lanczos algorithm for non-Hermitian matrices*, SIAM J. Sci. Comput., 14 (1993), pp. 137–158.

[80] R. W. FREUND AND N. M. NACHTIGAL, *QMR: a quasi-minimal residual algorithm for non-hermitian linear systems*, Numer. Math., 60 (1991), pp. 315–339.

[81] ———, *An implementation of the QMR method based on coupled two-term recurrences*, SIAM J. Sci. Comput., 15 (1994), pp. 313–337.

[82] D. M. GAY, *Some convergence properties of Broyden's method*, SIAM J. Numer. Anal., 16 (1979), pp. 623–630.

[83] C. W. GEAR, *Numerical Initial Value Problems in Ordinary Differential Equations*, Prentice-Hall, Englewood Cliffs, NJ, 1971.

[84] R. R. GERBER AND F. T. LUK, *A generalized Broyden's method for solving simultaneous linear equations*, SIAM J. Numer. Anal., 18 (1981), pp. 882–890.

[85] P. E. GILL AND W. MURRAY, *Quasi-Newton methods for unconstrained optimization*, J. I. M. A., 9 (1972), pp. 91–108.

[86] ———, *Safeguarded steplength algorithms for optimization using descent methods*, Tech. Report NAC 37, National Physical Laboratory Report, Teddington, England, 1974.

[87] P. E. GILL, W. MURRAY, AND M. H. WRIGHT, *Practical Optimization*, Academic Press, London, 1981.

[88] A. A. GOLDSTEIN, *Constructive Real Analysis*, Harper and Row, New York, 1967.

[89] G. H. GOLUB AND C. G. VANLOAN, *Matrix Computations*, Johns Hopkins University Press, Baltimore, 1983.

[90] A. GRIEWANK, *Analysis and modification of Newton's method at singularities*, PhD thesis, Australian National University, 1981.

[91] ———, *Rates of convergence for secant methods on nonlinear problems in Hilbert space*, in Numerical Analysis, Proceedings Guanajuato , Mexico 1984, Lecture Notes in Math., No, 1230, J. P. Hennart, ed., Springer-Verlag, Heidelberg, 1986, pp. 138–157.

[92] ———, *The solution of boundary value problems by Broyden based secant methods*, in Computational Techniques and Applications: CTAC 85, Proceedings of CTAC, Melbourne, August 1985, J. Noye and R. May, eds., North Holland, Amsterdam, 1986, pp. 309–321.

[93] ———, *On the iterative solution of differential and integral equations using secant updating techniques*, in The State of the Art in Numerical Analysis, A. Iserles and M. J. D. Powell, eds., Clarendon Press, Oxford, 1987, pp. 299–324.

[94] A. GRIEWANK AND G. F. CORLISS, eds., *Automatic Differentiation of Algorithms: Theory, Implementation, and Application*, Society for Industrial and Applied Mathematics, Philadelphia, PA, 1991.

[95] A. GRIEWANK AND P. L. TOINT, *Local convergence analysis for partitioned quasi-newton updates*, Numer. Math., 39 (1982), pp. 429–448.

[96] ———, *Partitioned variable metric methods for large sparse optimization problems*, Numer. Math., 39 (1982), pp. 119–137.

[97] W. A. GRUVER AND E. SACHS, *Algorithmic Methods In Optimal Control*, Pitman, London, 1980.

[98] M. H. GUTKNECHT, *Variants of BICGSTAB for matrices with complex spectrum*, SIAM J. Sci. Comput., 14 (1993), pp. 1020–1033.

[99] W. HACKBUSCH, *Multi-Grid Methods and Applications*, vol. 4 of Springer Series in Computational Mathematics, Springer-Verlag, New York, 1985.

[100] ——, *Multigrid methods of the second kind*, in Multigrid Methods for Integral and Differential Equations, Oxford University Press, Oxford, 1985.

[101] W. E. HART AND S. O. W. SOUL, *Quasi-Newton methods for discretized nonlinear boundary problems*, Journal of the Institute of Applied Mathematics, 11 (1973), pp. 351–359.

[102] H. V. HENDERSON AND S. R. SEARLE, *On deriving the inverse of a sum of matrices*, SIAM Rev., 23 (1981), pp. 53–60.

[103] M. R. HESTENES AND E. STEIFEL, *Methods of conjugate gradient for solving linear systems*, J. of Res. Nat. Bureau Standards, 49 (1952), pp. 409–436.

[104] D. M. HWANG AND C. T. KELLEY, *Convergence of Broyden's method in Banach spaces*, SIAM J. Optim., 2 (1992), pp. 505–532.

[105] E. ISAACSON AND H. B. KELLER, *Analysis of numerical methods*, John Wiley, New York, 1966.

[106] L. KANTOROVICH AND G. AKILOV, *Functional Analysis*, 2nd ed., Pergamon Press, New York, 1982.

[107] L. V. KANTOROVICH, *Functional analysis and applied mathematics*, Uspehi Mat. Nauk., 3 (1948), pp. 89–185. translation by C. Benster as Nat. Bur. Standards Report 1509. Washington, D. C., 1952.

[108] H. B. KELLER, *Newton's method under mild differentiability conditions*, J. Comput. Sys. Sci, 4 (1970), pp. 15–28.

[109] ——, *Lectures on Numerical Methods in Bifurcation Theory*, Tata Institute of Fundamental Research, Lectures on Mathematics and Physics, Springer-Verlag, New York, 1987.

[110] C. T. KELLEY, *Solution of the Chandrasekhar H-equation by Newton's method*, J. Math. Phys., 21 (1980), pp. 1625–1628.

[111] ——, *A fast multilevel algorithm for integral equations*, SIAM J. Numer. Anal., 32 (1995), pp. 501–513.

[112] C. T. KELLEY AND E. W. SACHS, *A quasi-Newton method for elliptic boundary value problems*, SIAM J. Numer. Anal., 24 (1987), pp. 516–531.

[113] ——, *A pointwise quasi-Newton method for unconstrained optimal control problems*, Numer. Math., 55 (1989), pp. 159–176.

[114] ——, *Fast algorithms for compact fixed point problems with inexact function evaluations*, SIAM J. Sci. Statist. Comput., 12 (1991), pp. 725–742.

[115] ——, *A new proof of superlinear convergence for Broyden's method in Hilbert space*, SIAM J. Optim., 1 (1991), pp. 146–150.

[116] ——, *Pointwise Broyden methods*, SIAM J. Optim., 3 (1993), pp. 423–441.

[117] ——, *Multilevel algorithms for constrained compact fixed point problems*, SIAM J. Sci. Comput., 15 (1994), pp. 645–667.

[118] C. T. KELLEY AND R. SURESH, *A new acceleration method for Newton's method at singular points*, SIAM J. Numer. Anal., 20 (1983), pp. 1001–1009.

[119] C. T. KELLEY AND Z. Q. XUE, *Inexact Newton methods for singular problems*, Optimization Methods and Software, 2 (1993), pp. 249–267.

[120] ——, *GMRES and integral operators*, SIAM J. Sci. Comput., 17 (1996), pp. 217–226.

[121] T. KERKHOVEN AND Y. SAAD, *On acceleration methods for coupled nonlinear elliptic systems*, Numerische Mathematik, 60 (1992), pp. 525–548.

[122] C. LANCZOS, *Solution of linear equations by minimized iterations*, J. Res. Natl. Bur. Stand., 49 (1952), pp. 33–53.

[123] T. A. MANTEUFFEL, *Adaptive procedure for estimating parameters for the nonsymmetric Tchebychev iteration*, Numer. Math., 31 (1978), pp. 183–208.

[124] T. A. MANTEUFFEL AND S. PARTER, *Preconditioning and boundary conditions*, SIAM J. Numer. Anal., 27 (1990), pp. 656–694.

[125] E. S. MARWIL, *Convergence results for Schubert's method for solving sparse nonlinear equations*, SIAM J. Numer. Anal., (1979), pp. 588–604.

[126] S. MCCORMICK, *Multilevel Adaptive Methods for Partial Differential Equations*, Society for Industrial and Applied Mathematics, Philadelphia, PA, 1989.

[127] J. A. MEIJERINK AND H. A. VAN DER VORST, *An iterative solution method for linear systems of which the coefficient matrix is a symmetric M-matrix*, Math. Comput., 31 (1977), pp. 148–162.

[128] C. D. MEYER, *Matrix Analysis and Applied Linear Algebra*, forthcoming.

[129] J. J. MORÉ, *Recent developments in algorithms and software for trust region methods*, in Mathematical Programming: The State of the Art, A. Bachem, M. Gröschel, and B. Korte, eds., Springer-Verlag, Berlin, 1983, pp. 258–287.

[130] J. J. MORÉ AND J. A. TRANGENSTEIN, *On the global convergence of Broyden's method*, Math. Comput., 30 (1976), pp. 523–540.

[131] T. E. MOTT, *Newton's method and multiple roots*, Amer. Math. Monthly, 64 (1957), pp. 635–638.

[132] T. W. MULLIKIN, *Some probability distributions for neutron transport in a half space*, J. Appl. Probab., 5 (1968), pp. 357–374.

[133] W. MURRAY AND M. L. OVERTON, *Steplength algorithms for minimizing a class of nondifferentiable functions*, Computing, 23 (1979), pp. 309–331.

[134] N. M. NACHTIGAL, S. C. REDDY, AND L. N. TREFETHEN, *How fast are nonsymmetric matrix iterations?*, SIAM J. Matrix Anal. Appl., 13 (1992), pp. 778–795.

[135] N. M. NACHTIGAL, L. REICHEL, AND L. N. TREFETHEN, *A hybrid gmres algorithm for nonsymmetric linear systems*, SIAM J. Matrix Anal. Appl., 13 (1992).

[136] S. G. NASH, *Preconditioning of truncated Newton methods*, SIAM J. Sci. Statist. Comput., 6 (1985), pp. 599–616.

[137] ———, *Who was Raphson? (an answer)*. Electronic Posting to NA-Digest, v92n23, 1992.

[138] J. L. NAZARETH, *Conjugate gradient methods less dependent on conjugacy*, SIAM Rev., 28 (1986), pp. 501–512.

[139] O. NEVANLINNA, *Convergence of Iterations for Linear Equations*, Birkhäuser, Basel, 1993.

[140] I. NEWTON, *The Mathematical Papers of Isaac Newton (7 volumes)*, D. T. Whiteside, ed., Cambridge University Press, Cambridge, 1967–1976.

[141] B. NOBLE, *Applied Linear Algebra*, Prentice Hall, Englewood Cliffs, NJ, 1969.

[142] J. NOCEDAL, *Theory of algorithms for unconstrained optimization*, Acta Numerica, 1 (1991), pp. 199–242.

[143] D. P. O'LEARY, *Why Broyden's nonsymmetric method terminates on linear equations*, SIAM J. Optim., 4 (1995), pp. 231–235.

[144] J. M. ORTEGA, *Numerical Analysis A Second Course*, Classics in Appl. Math., No. 3, Society for Industrial and Applied Mathematics, Philadelphia, PA, 1990.

[145] J. M. ORTEGA AND W. C. RHEINBOLDT, *Iterative Solution of Nonlinear Equations in Several Variables*, Academic Press, New York, 1970.

[146] A. M. OSTROWSKI, *Solution of Equations and Systems of Equations*, Academic

Press, New York, 1960.

[147] B. N. PARLETT, *The Symmetric Eigenvalue Problem*, Prentice Hall, Englewood Cliffs, NJ, 1980.

[148] B. N. PARLETT, D. R. TAYLOR, AND Z. A. LIU, *A look-ahead Lanczos algorithm for unsymmetric matrices*, Math. Comput., 44 (1985), pp. 105–124.

[149] D. W. PEACEMAN AND H. H. RACHFORD, *The numerical solution of parabolic and elliptic differential equations*, J. Soc. Indus. Appl. Math., 11 (1955), pp. 28–41.

[150] G. PETERS AND J. WILKINSON, *Inverse iteration, ill-conditioned equations and Newton's method*, SIAM Rev., 29 (1979), pp. 339–360.

[151] L. R. PETZOLD, *A description of DASSL: a differential/algebraic system solver*, in Scientific Computing, R. S. Stepleman et al., eds., North Holland, Amsterdam, 1983, pp. 65–68.

[152] E. PICARD, *Mémoire sur la théorie des équations aux dérivées partielles et la méthode des approximations successives*, J. de Math. ser 4, 6 (1890), pp. 145–210.

[153] M. J. D. POWELL, *A hybrid method for nonlinear equations*, in Numerical Methods for Nonlinear Algebraic Equations, Gordon and Breach, New York, 1970, pp. 87–114.

[154] ——, *Convergence properties of a class of minimization algorithms*, in Nonlinear Programming 2, O. L. Mangasarian, R. R. Meyer, and S. M. Robinson, eds., Academic Press, New York, 1975, pp. 1–27.

[155] L. B. RALL, *Convergence of the Newton process to multiple solutions*, Numer. Math., 9 (1961), pp. 23–37.

[156] J. RAPHSON, *Analysis aequationum universalis seu ad aequationes algebraicas resolvendas methodus generalis, et expedita, ex nova infinitarum serierum doctrina, deducta ac demonstrata*. Original in British Library, London, 1690.

[157] G. W. REDDIEN, *On Newton's method for singular problems*, SIAM J. Numer. Anal., 15 (1978), pp. 993–986.

[158] J. K. REID, *Least squares solution of sparse systems of nonlinear equations by a modified Marquardt algorithm*, in Proc. NATO Conf. at Cambridge, July 1972, North Holand, Amsterdam, 1973, pp. 437–445.

[159] W. C. RHEINBOLDT, *Numerical Analysis of Parametrized Nonlinear Equations*, John Wiley, New York, 1986.

[160] J. R. RICE, *Experiments on Gram-Schmidt orthogonalization*, Math. Comput., 20 (1966), pp. 325–328.

[161] T. J. RIVLIN, *The Chebyshev Polynomials*, John Wiley, New York, 1974.

[162] H. L. ROYDEN, *Real Analysis*, 2nd ed., Macmillan, New York, 1968.

[163] Y. SAAD, *Practical use of polynomial preconditionings for the conjugate gradient method*, SIAM J. Sci. Statist. Comput., 6 (1985), pp. 865–881.

[164] ——, *Least squares polynomials in the complex plane and their use for solving nonsymmetric linear systems*, SIAM J. Numer. Anal., 24 (1987), pp. 155–169.

[165] ——, *ILUT: A dual threshold incomplete LU factorization*, Tech. Report 92-38, Computer Science Department, University of Minnesota, 1992.

[166] ——, *A flexible inner-outer preconditioned GMRES algorithm*, SIAM J. Sci. Comput., (1993), pp. 461–469.

[167] Y. SAAD AND M. SCHULTZ, *GMRES a generalized minimal residual algorithm for solving nonsymmetric linear systems*, SIAM J. Sci. Statist. Comput., 7 (1986), pp. 856–869.

[168] E. SACHS, *Broyden's method in Hilbert space*, Math. Programming, 35 (1986), pp. 71–82.

[169] P. E. SAYLOR AND D. C. SMOLARSKI, *Implementation of an adaptive algorithm*

for Richardson's method, Linear Algebrra Appl., 154/156 (1991), pp. 615–646.

[170] R. B. SCHNABEL AND P. D. FRANK, *Tensor methods for nonlinear equations,* SIAM J. Numer. Anal., 21 (1984), pp. 815–843.

[171] E. SCHRÖDER, *Über unendlich viele Algorithmen zur Auflosung der Gleichungen,* Math. Ann., 2 (1870), pp. 317–365.

[172] L. K. SCHUBERT, *Modification of a quasi-Newton method for nonlinear equations with sparse Jacobian,* Math. Comput., 24 (1970), pp. 27–30.

[173] G. A. SCHULTZ, R. B. SCHNABEL, AND R. H. BYRD, *A family of trust-region-based algorithms for unconstrained minimization with strong global convergence properties,* SIAM J. Numer. Anal., 22 (1985), pp. 47–67.

[174] V. E. SHAMANSKII, *A modification of Newton's method,* Ukran. Mat. Zh., 19 (1967), pp. 133–138. (In Russian.)

[175] A. H. SHERMAN, *On Newton-iterative methods for the solution of systems of nonlinear equations,* SIAM J. Numer. Anal., 14 (1978), pp. 755–774.

[176] J. SHERMAN AND W. J. MORRISON, *Adjustment of an inverse matrix corresponding to changes in the elements of a given column or a given row of the original matrix (abstract),* Ann. Math. Statist., 20 (1949), p. 621.

[177] ——, *Adjustment of an inverse matrix corresponding to a change in one element of a given matrix,* Ann. Math. Statist., 21 (1950), pp. 124–127.

[178] K. SIGMON, *MATLAB Primer, Fourth Edition,* CRC Press, Boca Raton, FL, 1994.

[179] D. C. SMOLARSKI AND P. E. SAYLOR, *An optimum iterative method for solving any linear system with a square matrix,* BIT, 28 (1988), pp. 163–178.

[180] P. SONNEVELD, *CGS, a fast Lanczos-type solver for nonsymmetric linear systems,* SIAM J. Sci. Statist. Comput., 10 (1989), pp. 36–52.

[181] D. C. SORENSEN, *Newton's method with a model trust region modification,* SIAM J. Numer. Anal., 19 (1982), pp. 409–426.

[182] G. STARKE, *Alternating direction preconditioning for nonsymmetric systems of linear equations,* SIAM J. Sci. Comput., 15 (1994), pp. 369–385.

[183] G. STARKE AND R. S. VARGA, *A hybrid Arnoldi-Faber iterative method for nonsymmetric systems of linear equations,* Numer. Math., 64 (1993), pp. 213–239.

[184] G. W. STEWART, *Introduction to matrix computations,* Academic Press, New York, 1973.

[185] E. L. STIEFEL, *Kernel polynomials in linear algebra and their numerical applications,* U.S. National Bureau of Standards, Applied Mathematics Series, 49 (1958), pp. 1–22.

[186] P. N. SWARZTRAUBER, *The methods of cyclic reduction, Fourier analysis and the FACR algorithm for the discrete solution of Poisson's equation on a rectangle,* SIAM Rev., 19 (1977), pp. 490–501.

[187] ——, *Approximate cyclic reduction for solving Poisson's equation,* SIAM J. Sci. Statist. Comput., 8 (1987), pp. 199–209.

[188] P. N. SWARZTRAUBER AND R. A. SWEET, *Algorithm 541: Efficient FORTRAN subprograms for the solution of elliptic partial differential equations,* ACM Trans. Math. Software, 5 (1979), pp. 352–364.

[189] P. L. TOINT, *On sparse and symmetric matrix updating subject to a linear equation,* Math. Comput., 31 (1977), pp. 954–961.

[190] J. F. TRAUB, *Iterative Methods for the Solution of Equations,* Prentice Hall, Englewood Cliffs, NJ, 1964.

[191] H. A. VAN DER VORST, *Bi-CGSTAB: A fast and smoothly converging variant to Bi-CG for the solution of nonlinear systems,* SIAM J. Sci. Statist. Comput., 13

(1992), pp. 631–644.

[192] H. A. VAN DER VORST AND C. VUIK, *The superlinear convergence behaviour of GMRES*, Journal Comput. Appl. Math., 48 (1993), pp. 327–341.

[193] R. S. VARGA, *Matrix Iterative Analysis*, Prentice Hall, Englewood Cliffs, NJ, 1962.

[194] E. L. WACHSPRESS, *Iterative Solution of Elliptic Systems and Applications to the Neutron Diffusion Equations of Reactor Physics*, Prentice Hall, Englewood Cliffs, NJ, 1966.

[195] H. F. WALKER, *Implementation of the GMRES method using Householder transformations*, SIAM J. Sci. Statist. Comput., 9 (1989), pp. 815–825.

[196] ——, *Residual smoothing and peak/plateau behavior in Krylov subspace methods*, Applied Numer. Math., 19 (1995), pp. 279–286.

[197] H. F. WALKER AND L. ZHOU, *A simpler GMRES*, J. Numer. Lin. Alg. Appl., 6 (1994), pp. 571–581.

[198] L. T. WATSON, S. C. BILLUPS, AND A. P. MORGAN, *Algorithm 652: HOMPACK: A suite of codes for globally convergent homotopy algorithms*, ACM Trans. Math. Software, 13 (1987), pp. 281–310.

[199] M. A. WOODBURY, *Inverting modified matrices*, Memorandum Report 42, Statistical Research Group, Princeton NJ, 1950.

[200] D. M. YOUNG, *Iterative Solution of Large Linear Systems*, Academic Press, New York, 1971.

[201] T. J. YPMA, *The effect of rounding errors on Newton-like methods*, IMA J. Numer. Anal., 3 (1983), pp. 109–118.

[202] L. ZHOU AND H. F. WALKER, *Residual smoothing techniques for iterative methods*, SIAM J. Sci. Comput., 15 (1994), pp. 297–312.

Index